METHODS IN MOLECULAR BIOLOGY

For further volumes:
http://www.springer.com/series/7651

For over 35 years, biological scientists have come to rely on the research protocols and methodologies in the critically acclaimed *Methods in Molecular Biology* series. The series was the first to introduce the step-by-step protocols approach that has become the standard in all biomedical protocol publishing. Each protocol is provided in readily-reproducible step-by-step fashion, opening with an introductory overview, a list of the materials and reagents needed to complete the experiment, and followed by a detailed procedure that is supported with a helpful notes section offering tips and tricks of the trade as well as troubleshooting advice. These hallmark features were introduced by series editor Dr. John Walker and constitute the key ingredient in each and every volume of the *Methods in Molecular Biology* series. Tested and trusted, comprehensive and reliable, all protocols from the series are indexed in PubMed.

Fusobacterium nucleatum

Methods and Protocols

Edited by

Hung Ton-That

School of Dentistry, University of California - Los Angeles, Los Angeles, CA, USA

Humana Press

Editor
Hung Ton-That
School of Dentistry
University of California - Los Angeles
Los Angeles, CA, USA

ISSN 1064-3745 ISSN 1940-6029 (electronic)
Methods in Molecular Biology
ISBN 978-1-0716-5387-6 ISBN 978-1-0716-5388-3 (eBook)
https://doi.org/10.1007/978-1-0716-5388-3

This Humana imprint is published by the registered company Springer Science+Business Media, LLC, part of Springer Nature.
The registered company address is: 1 New York Plaza, New York, NY 10004, U.S.A.

Preface

Fusobacterium nucleatum: Methods and Protocols presents a compendium of both classic and modern experimental methods designed to illuminate the biology, pathogenicity, and host interactions of *Fusobacterium nucleatum*. Research on *F. nucleatum* has evolved substantially. Since its initial description in the early 1900s as an elongated, spindle-shaped anaerobe inhabiting the oral cavity and gingival crevices, *F. nucleatum* was long regarded as a relatively benign member of the oral microbiota, primarily associated with plaque formation and periodontal disease. Pioneering microbiological studies during the mid- to late twentieth century, however, revealed its central role as an organizer of dental biofilms, distinguished by its robust adhesion, coaggregation, and immunomodulatory capabilities. These foundational insights laid the groundwork for contemporary explorations into its biological complexity.

A major paradigm shift in understanding the organism's role as a pathobiont has emerged in the past decade, when clinical genomic and microbiome investigations linked *F. nucleatum* to colorectal cancer (CRC) and a range of other systemic diseases. This discovery broadened the significance of the organism far beyond oral ecology, positioning it as a critical contributor to cancer biology, inflammation research, and host–microbe interactions. Advances in sequencing, imaging, metabolomics, and genetic manipulation have propelled *F. nucleatum* research into a vibrant, multidisciplinary field. Yet due to its strict anaerobic physiology, the species remains technically challenging to manipulate for various in vitro and in vivo studies, underscoring the need for a centralized, standardized resource of reliable and accessible experimental protocols.

This volume was assembled to address that need. Authored by leading experts, the chapters provide practical guidance spanning microbiology, molecular biology, imaging, genetic engineering, metabolomics, and in vivo modeling. Together, these methods offer a foundational toolkit for both new and experienced researchers seeking to investigate *F. nucleatum* virulence mechanisms, colonization strategies, and its increasingly recognized contributions to diseases such as CRC, oral squamous cell carcinoma, and periodontitis.

The volume opens with methods for visualizing and isolating *F. nucleatum* within its native oral environment, including CLASI-FISH analysis of oral biofilms and protocols for subspecies isolation and identification from clinical specimens. Advances in genetic manipulation—historically a major bottleneck in *Fusobacterium* research—are featured prominently, including enhanced transformation approaches, conditional plasmid systems for markerless gene deletion, CRISPR interference strategies, and broader frameworks for genetic engineering across *Fusobacterium* species. Additional chapters highlight essential cellular processes such as surface protein assembly and peptidoglycan biosynthesis, as well as the isolation and characterization of molecular structures including exosomal sRNAs, primary RNA for dRNA-seq, bacterial microcompartments, and outer membrane vesicles.

Bridging molecular analysis with host interactions, the book further presents protocols for asymmetric coculture systems, 3D tumor spheroid models, untargeted metabolite profiling of infected mammalian cells, and several in vivo platforms—from murine models of periodontitis, oral cancer, and CRC to a *Caenorhabditis elegans* infection model.

Together, these chapters illustrate the diverse experimental contexts in which *F. nucleatum* exerts its influence and provide the methodological infrastructure needed to explore its biological impact across systems.

This comprehensive methodological foundation reflects the collective efforts of numerous researchers across disciplines. I am therefore deeply grateful to all contributing authors for their expertise, enthusiasm, and commitment to advancing the study of *F. nucleatum*. I extend my heartfelt thanks to my longtime collaborator and friend, Asis Das, for his insights, support, and many thoughtful contributions over the years, which have greatly enriched our work and this volume. I am equally indebted to my former and current lab members, whose creativity, hard work, and dedication have been intrumental to advancing *F. nucleatum* genetic and molecular research. I also wish to thank my son, HyLam Ton-That, for his invaluable assistance in editing the chapters; his keen attention to detail strengthened this volume in countless ways. Finally, my sincere gratitude goes to John Walker, the series editor, whose guidance, support, and steady encouragement made this book possible.

It is my hope that this collection serves as a valuable resource for the scientific community, inspires new lines of investigation, and catalyzes discoveries that deepen our understanding of *F. nucleatum* biology and its implications for human health and disease—ultimately advancing the development of novel therapeutic strategies for wide-ranging systemic disorders.

Los Angeles, CA, USA ***Hung Ton-That***

Contents

Contributors

ANGELA AGNEW • *Molecular Biology Institute, University of California – Los Angeles, Los Angeles, CA, USA; Department of Microbiology, Immunology & Molecular Genetics, University of California - Los Angeles, Los Angeles, CA, USA*

SUHANI B. BHAKTA • *Molecular Biology Institute, University of California – Los Angeles, Los Angeles, CA, USA; Department of Microbiology, Immunology & Molecular Genetics, University of California – Los Angeles, Los Angeles, CA, USA*

G C BIBEK • *Department of Microbiology & Molecular Genetics, The University of Texas Health Science Center, Houston, TX, USA*

CHRISTINA BORLAND • *Division of Biomaterial and Biomedical Sciences, School of Dentistry, Oregon Health & Science University (OHSU), Portland, OR, USA; Department of Molecular Microbiology and Immunology, Oregon Health & Science University (OHSU), Portland, OR, USA*

TIMMIE A. BRITTON • *Division of Gastroenterology and Hepatology, Mayo Clinic, Rochester, MN, USA; Molecular Biology Institute, University of California – Los Angeles, Los Angeles, CA, USA*

NICOLE M. CADY • *Department of Microbiology and Immunology, University of Michigan Medical School, Ann Arbor, MI, USA*

LUJIA CEN • *Department of Microbiology, The ADA Forsyth Institute, Cambridge, MA, USA*

CHUNGYU CHANG • *Division of Oral & Systemic Health Sciences, School of Dentistry, University of California – Los Angeles, Los Angeles, CA, USA*

JIA CHANG • *Division of Regenerative and Reconstructive Science, School of Dentistry, University of California - Los Angeles, Los Angeles, CA, USA*

YUNG-FU CHANG • *Department of Population Medicine and Diagnostic Sciences, Cornell University College of Veterinary Medicine, Ithaca, NY, USA*

QIUYING CHEN • *Department of Pharmacology, Weill Cornell Medical College, New York, NY, USA*

YIMIN CHEN • *Division of Oral and Systemic Health Sciences, School of Dentistry, University of California - Los Angeles, Los Angeles, CA, USA*

YI-WEI CHEN • *Division of Oral and Systemic Health Sciences, School of Dentistry, University of California - Los Angeles, Los Angeles, CA, USA*

HUIYONG CHENG • *Department of Pharmacology, Weill Cornell Medical College, New York, NY, USA*

COLLINS I. CHIMEZIE • *Program in Chemical Biology, Life Sciences Institute, University of Michigan, Ann Arbor, MI, USA*

JOHN CROW • *Molecular Biology Institute, University of California – Los Angeles, Los Angeles, CA, USA*

ASIS DAS • *Department of Medicine, Neag Comprehensive Cancer Center, School of Medicine, University of Connecticut Health Center, Farmington, CT, USA*

NICHOLAS DE MOJANA DI COLOGNA • *Department of Biological Sciences and RNA Institute, University at Albany, SUNY, Albany, NY, USA*

DANA FRANKLIN • *Molecular Biology Institute, University of California – Los Angeles, Los Angeles, CA, USA*

RUTIKA GAVATE • *Genomic Medicine, MD Anderson Cancer Center, Houston, TX, USA*

STEVEN S. GROSS • *Department of Pharmacology, Weill Cornell Medical College, New York, NY, USA*

AMBER E. HALEY • *Program in Chemical Biology, Life Sciences Institute, University of Michigan, Ann Arbor, MI, USA*

BING HAN • *Division of Oral and Systemic Health Sciences, School of Dentistry, University of California - Los Angeles, Los Angeles, CA, USA; Laboratory of Molecular Signaling, Jonsson Comprehensive Cancer Center, University of California - Los Angeles, Los Angeles, CA, USA*

XUESONG HE • *Department of Microbiology, The ADA Forsyth Institute, Cambridge, MA, USA*

YVONNE L. HERNANDEZ-KAPILA • *Section of Biosystems and Function, School of Dentistry, University of California Los Angeles, Los Angeles, CA, USA*

MEREDITH A. J. HULLAR • *Cancer Prevention, Public Health Sciences Division, Fred Hutchinson Cancer Center, Seattle, WA, USA*

CHRISTOPHER D. JOHNSTON • *Vaccine and Infectious Disease Division, Fred Hutchinson Cancer Center, Seattle, WA, USA; Genomic Medicine, MD Anderson Cancer Center, Houston, TX, USA*

ORSALEM KAHSAI • *Cancer Prevention, Public Health Sciences Division, Fred Hutchinson Cancer Center, Seattle, WA, USA*

PACHIYAPPAN KAMARAJAN • *Section of Biosystems and Function, School of Dentistry, University of California Los Angeles, Los Angeles, CA, USA*

OMRI KARIV • *Department of Microbiology, Immunology and Molecular Genetics, University of California – Los Angeles, Los Angeles, CA, USA*

REUBEN H. KIM • *The Shapiro Family Laboratory of Viral Oncology and Aging Research, Section of Restorative Dentistry, UCLA School of Dentistry, Los Angeles, CA, USA; Division of Preventive and Restorative Sciences, School of Dentistry, University of California – Los Angeles, Los Angeles, CA, USA; UCLA Jonsson Comprehensive Cancer Center, Los Angeles, CA, USA*

SHARON KIM • *The Shapiro Family Laboratory of Viral Oncology and Aging Research, Section of Restorative Dentistry, UCLA School of Dentistry, Los Angeles, CA, USA; Division of Preventive and Restorative Sciences, School of Dentistry, University of California – Los Angeles, Los Angeles, CA, USA*

MADELINE KRIEGER • *Cancer Early Detection Advanced Research Center, Knight Cancer Institute, Oregon Health & Science University (OHSU), Portland, OR, USA*

AJAY KUMAR • *The ADA Forsyth Institute, Department of Microbiology, Somerville, MA, USA*

ALEX A. LEMUS • *Department of Biological Sciences and RNA Institute, University at Albany, SUNY, Albany, NY, USA*

KAITLYN N. LEWIS • *Genomic Medicine, MD Anderson Cancer Center, Houston, TX, USA*

JIAHE LI • *Department of Biomedical Engineering, College of Engineering and School of Medicine, University of Michigan, Ann Arbor, MI, USA*

JINJING LIU • *Department of Population Medicine and Diagnostic Sciences, Cornell University College of Veterinary Medicine, Ithaca, NY, USA*

LING LIU • *Department of Laboratory Medicine, Key Laboratory of Precision Medicine for Viral Diseases, Guangxi Health Commission Key Laboratory of Clinical Biotechnology, Liuzhou People's Hospital, Liu Zhou, China*

WEI LIU • *Division of Oral and Systemic Health Sciences, School of Dentistry, University of California - Los Angeles, Los Angeles, CA, USA; Laboratory of Molecular Signaling, Jonsson Comprehensive Cancer Center, University of California - Los Angeles, Los Angeles, CA, USA*

XUE LIU • *Guangdong Provincial Key Laboratory of Regional Immunity and Diseases, Department of Pathogen Biology, International Cancer Center, Shenzhen University Medical School, Shenzhen, China*

ELSA F. MCMAHON • *Vaccine and Infectious Disease Division, Fred Hutchinson Cancer Center, Seattle, WA, USA*

JUSTIN MERRITT • *Division of Biomaterial and Biomedical Sciences, School of Dentistry, Oregon Health & Science University (OHSU), Portland, OR, USA; Department of Molecular Microbiology and Immunology, Oregon Health & Science University (OHSU), Portland, OR, USA*

POLLY A. NEWCOMB • *Cancer Prevention, Public Health Sciences Division, Fred Hutchinson Cancer Center, Seattle, WA, USA*

ZEYANG PANG • *Department of Biomedical Engineering, College of Engineering and School of Medicine, University of Michigan, Ann Arbor, MI, USA*

NO-HEE PARK • *The Shapiro Family Laboratory of Viral Oncology and Aging Research, Section of Restorative Dentistry, UCLA School of Dentistry, Los Angeles, CA, USA; UCLA Jonsson Comprehensive Cancer Center, Los Angeles, CA, USA*

AMANDA I. PHIPPS • *Epidemiology, Public Health Sciences Division, Fred Hutchinson Cancer Center, Seattle, WA, USA; Department of Epidemiology, University of Washington, Seattle, WA, USA*

FALK PONATH • *Department of Immunology, The University of Texas MD Anderson Cancer Center, Houston, TX, USA*

ALLAN RADAIC • *Section of Biosystems and Function, School of Dentistry, University of California Los Angeles, Los Angeles, CA, USA*

BRYAN SANCHEZ • *Molecular Biology Institute, University of California – Los Angeles, Los Angeles, CA, USA; Medical Scientist Training Program, University of California – Los Angeles, Los Angeles, CA, USA*

THOMAS M. SCHMIDT • *Department of Microbiology and Immunology, University of Michigan Medical School, Ann Arbor, MI, USA; Department of Internal Medicine, University of Michigan, Ann Arbor, MI, USA; Department of Ecology and Evolutionary Biology, University of Michigan, Ann Arbor, MI, USA*

KEXIN TAN • *Department of Microbiology & Molecular Genetics, The University of Texas Health Science Center, Houston, TX, USA*

HUNG TON-THAT • *Molecular Biology Institute, University of California – Los Angeles, Los Angeles, CA, USA; Department of Microbiology, Immunology & Molecular Genetics, University of California - Los Angeles, Los Angeles, CA, USA; California NanoSystems Institute, University of California – Los Angeles, Los Angeles, CA, USA; Division of Oral & Systemic Health Sciences, School of Dentistry, University of California - Los Angeles, Los Angeles, CA, USA*

ALEX M. VALM • *Department of Biological Sciences and RNA Institute, University at Albany, SUNY, Albany, NY, USA*

JÖRG VOGEL • *Institute of Molecular Infection Biology (IMIB), University of Würzburg, Würzburg, Germany; University of Würzburg, Würzburg, Germany; Helmholtz Institute for RNA-based Infection Research (HIRI), Helmholtz Centre for Infection Research (HZI), Würzburg, Germany*

CUN-YU WANG • *Division of Oral and Systemic Health Sciences, School of Dentistry, University of California - Los Angeles, Los Angeles, CA, USA; Laboratory of Molecular Signaling, Jonsson Comprehensive Cancer Center, University of California - Los Angeles, Los Angeles, CA, USA*
RUOGU WANG • *Department of Biological Sciences and RNA Institute, University at Albany, SUNY, Albany, NY, USA*
CHENGGANG WU • *Department of Microbiology & Molecular Genetics, The University of Texas Health Science Center, Houston, TX, USA*
SHIQI XU • *Department of Microbiology & Molecular Genetics, The University of Texas Health Science Center, Houston, TX, USA*
MARTHA A. ZEPEDA-RIVERA • *Vaccine and Infectious Disease Division, Fred Hutchinson Cancer Center, Seattle, WA, USA; Genomic Medicine, MD Anderson Cancer Center, Houston, TX, USA*
YU ZHANG • *Guangdong Provincial Key Laboratory of Regional Immunity and Diseases, Department of Pathogen Biology, International Cancer Center, Shenzhen University Medical School, Shenzhen, China*
Z. HONG ZHOU • *Molecular Biology Institute, University of California – Los Angeles, Los Angeles, CA, USA; Department of Microbiology, Immunology & Molecular Genetics, University of California - Los Angeles, Los Angeles, CA, USA; California NanoSystems Institute, University of California – Los Angeles, Los Angeles, CA, USA*

Chapter 1

CLASI-FISH Imaging of *Fusobacterium* in Oral Microbial Communities

Nicholas de Mojana di Cologna, Ruogu Wang, Alex A. Lemus, and Alex M. Valm

Abstract

The human oral microbiome comprises a complex community of over 700 species of bacteria and other microbes with defined structure-function relationships, especially in dental plaque biofilms. *Fusobacterium nucleatum* is a ubiquitous organism in dental plaque known to coaggregate with multiple different species of oral microbes and hypothesized to play a crucial structural role in bridging early and late colonizing species in dental plaque. Microbial fluorescence in situ hybridization allows the mapping of bacteria in dense polymicrobial communities with exquisite taxonomic specificity. Here we provide an experimental procedure to label and visualize *Fusobacterium nucleatum* within dental plaque biofilms using combinatorial labeling and spectral imaging fluorescence in situ hybridization (CLASI-FISH). CLASI-FISH allows mapping of dozens of different microbial taxa in a single experiment. This streamlined protocol enhances accessibility for studying oral microbiomes. With rational probe design, this technique could be applied to map any organism of interest in the context of dozens of other microbial community members in diverse microbiomes.

Key words Fusobacterium, Multispectral imaging, Fluorescence imaging, CLASI-FISH

1 Introduction

1.1 CLASI-FISH for Studying Oral Biofilm Structure

Microscopy has revolutionized our understanding of the microbial world, from Antonie van Leeuwenhoek's earliest observations of bacteria to the development of super-resolution imaging techniques to map the molecular architecture within microbes [1–3]. Fluorescence in situ hybridization (FISH) has transformed visual identification of microbes by enabling specific fluorescent labeling of ribosomal RNA within intact cells, usually the 16S rRNA [4]. The 16S ribosomal RNA is a particularly well-suited molecular target for taxonomic identification of organisms with imaging due to its ubiquitous expression, high abundance in the cell cytoplasm leading to natural amplification of fluorescent signal,

Hung Ton-That (ed.), *Fusobacterium nucleatum: Methods and Protocols*, Methods in Molecular Biology, vol. 3055,
https://doi.org/10.1007/978-1-0716-5388-3_1,

and the presence of hypervariable sequence regions that differentiate bacterial species and serve as targets for taxon-specific FISH probes [5–7].

The Human Oral Microbiome Database (HOMD) comprises a well-curated collection of full-length 16S sequences from over 700 species of bacteria recognized to inhabit the human oral cavity [8]. Together with bioinformatics software tools, these computational resources allow for the fully in silico design and specificity testing of FISH probes for any species in the human oral microbiome [9]. While the number of probes that can be designed, in principle, to distinguish different species is enormous, in practice, the discrimination of more than a handful of labels is severely hampered by the broad excitation and emission spectra of biocompatible fluorophores. This results in significant cross-talk and bleed-through when imaging with filter cubes and noise in the recorded images [10]. Combinatorial labeling and spectral imaging fluorescence in situ hybridization (CLASI-FISH) overcomes this limitation, enabling highly multiplexed, system-level analysis of complex microbial community structure [11, 12]. CLASI-FISH extends the capabilities of traditional FISH by using combinations of fluorescent probes to label individual taxa. Combined with spectral imaging using commercially available confocal microscopes, which avoid the use of fluorescent filter cubes and capture the full emitted spectra of fluorophores, the technique has allowed simultaneous imaging of 15 or more taxa in the oral microbiome.

1.2 The Structural Role of Fusobacterium in the Dental Plaque

Oral biofilms are complex, highly organized structures in which bacterial interactions shape the organization and function of the community. A challenge for highly multiplexed FISH is the extraordinary heterogeneity in cell wall composition among oral bacteria. For example, gram-positive organisms, including *Streptococcus* species, are intermixed with gram-negative organisms, including *Porphyromonas* and *Fusobacterium* species [13]. When targeting a single species with FISH, partial enzymatic digestion of the cell wall can be achieved with, e.g., lysozyme to facilitate probe penetration of the cell; however, in mixed communities with multiplex FISH, no single permeabilization protocol will result in optimal labeling of all species [5]. The CLASI-FISH protocol described here eschews the use of enzymatic pre-treatment so as not to over-digest gram-negative organisms, e.g., *Fusobacterium* spp., with their relatively thin peptidoglycan layers, and instead employs an extended hybridization time (12–18 h) to attempt to optimize probe penetration and hybridization equilibration among structurally diverse bacteria.

Fusobacterium nucleatum is present in supragingival and subgingival plaque, and its relative abundance is positively correlated with periodontal disease [14]. Early in vitro work revealed that *F. nucleatum* can coaggregate with many other oral microbiome species, including both early and late colonizers [15]. Extensive molecular, genetic, and reduced complexity imaging analyses have identified several cell surface proteins involved in *Fusobacterium* coaggregation with other species, including RadD (*Streptococcus gordonii*, *Streptococcus mutans*, *Actinomyces naeslundii*), Fap2 (*Porphyromonas gingivalis*), FomA (*P. gingivalis*), and CmpA (*Streptococcus gordonii*), among others [16–21]. CLASI-FISH imaging of oral biofilms revealed that *Fusobacterium* spp. coaggregate with a small number of other species in mature supragingival plaque and that *Fusobacterium* cells localize to the middle layer of intact biofilms where they colocalize with several other bacterial filaments [11, 22]. Our own CLASI-FISH imaging of in vitro biofilms demonstrates that *Fusobacterium* cells make direct contact with multiple species in early biofilms, including *Streptococcus* and *Porphyromonas* (*see* Fig. 1). Together, these extensive observations support a model in which *Fusobacterium* plays a dynamic role in structuring dental plaque communities during ecological succession and biofilm development.

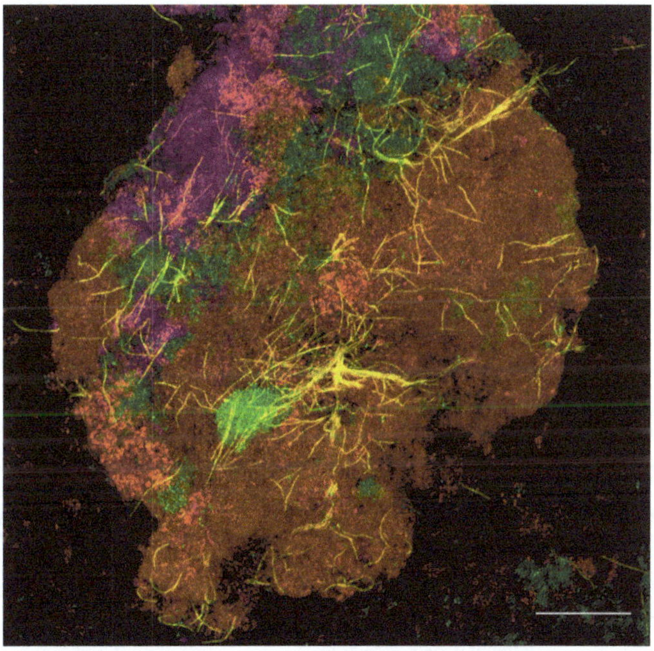

Fig. 1 CLASI-FISH image of an in vitro-grown biofilm derived from human dental plaque. Six taxon-specific probes were used, labeled with the following fluorophores: Alexa Fluor 488 (*Porphyromonas* sp., cyan), Alexa Fluor 514 (*Veillonella* sp., green), Alexa Fluor 555 (*Fusobacterium* sp., yellow), Rhodamine Red-X (*Pasteurellaceae*, orange), Alexa Fluor 594 (*Streptococcus* sp., red), and Alexa Fluor 633 (*Prevotella* sp., pink). Scale bar = 20 μm

1.3 Unmixing Fluorescence Signals in Spectral Images of Biological Samples

Multispectral microscopy allows for capturing the full emitted spectra of fluorescent labels in biological specimens, usually in discrete wavelength bands generated by prisms or diffraction gratings. Unmixing is the term for the computational analysis of multispectral fluorescence images for deconvolving the overlapping emission signals generated when fluorophores with highly overlapping spectra are used in a single experimental setup [23]. The most widely used method is least squares unmixing, which estimates fluorophore abundances by fitting a linear combination of known reference spectra to the measured pixel intensities in the unknown image.

This task conceptually resembles spectral unmixing in hyperspectral remote sensing, where mixed pixels are analyzed to identify the underlying materials [24–26]. The remote sensing field has developed many unmixing algorithms based on linear mixture models, incorporating statistical, geometric, or sparse priors [27–31].

The CLASI-FISH method includes reference images of labeled *Escherichia coli* containing only one fluorophore per sample. These reference images provide accurate spectral profiles for unmixing and serve as training data for supervised spectral classification. The availability of such prior information is a unique advantage of biological imaging and can be leveraged to improve the robustness of the unmixing process.

Specialized methods have been developed to address the challenges of biological multispectral imaging better. To model the Poisson-distributed photon counts, Wang et al. (2023) developed a sparse and low-rank Poisson regression approach that enhances unmixing in highly multiplexed datasets, including dental plaque biofilms [32]. Figure 2 shows a comparison between least squares unmixing and the sparse and low-rank Poisson regression method (SL-PRU) on a multiplex labeled dental plaque smear. SL-PRU produced cleaner cell boundaries and reduced background noise compared to linear least squares and notably improved the resolution of cells in the image. *Fusobacterium* cells appear as filamentous cells even in this low-resolution image. These results highlight the importance of developing unmixing models tailored to the unique characteristics of biological fluorescence imaging.

In addition to these methodological advances, future integration of morphological information offers promising avenues to further improve unmixing accuracy, particularly for distinguishing organisms with highly overlapping spectra. Many bacteria possess distinct shapes that can serve as valuable cues—for example, *Fusobacterium* exhibits a characteristic fusiform morphology, while other taxa may appear as cocci, rods, or spirochetes [33]. Combining spectral signatures with morphological features has the potential to enhance specificity in microbial identification, especially in densely packed or compositionally complex samples. Such

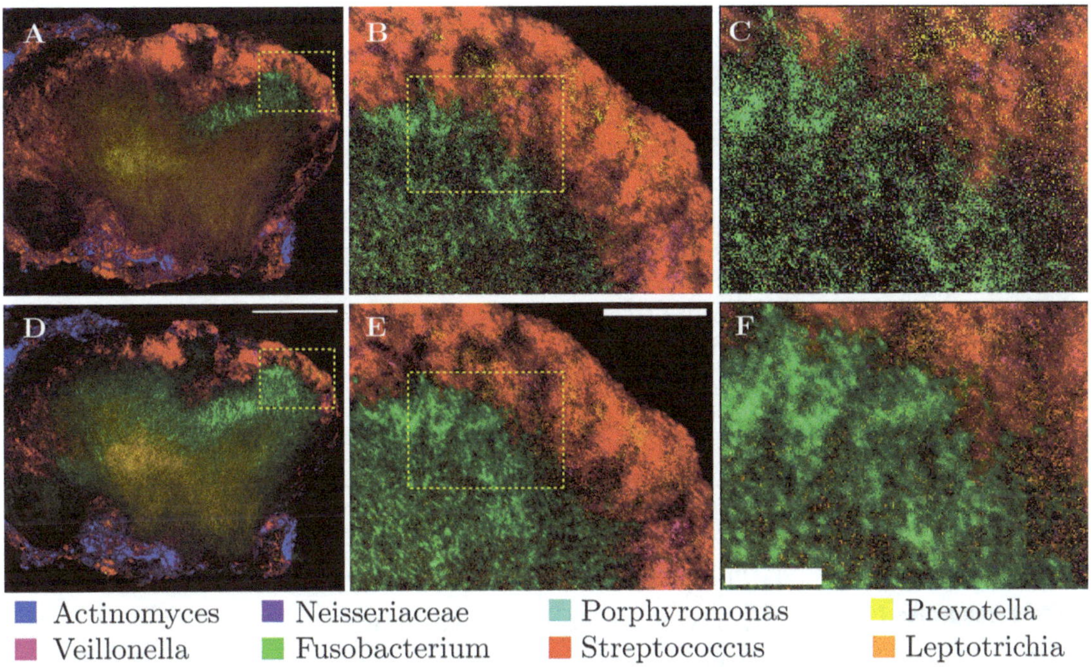

Fig. 2 Qualitative comparison of least squares and SL-PRU unmixing on a real biological sample. (**a–f**) Multispectral image of a dental plaque smear hedgehog structure after (**a–c**) least squares unmixing and (**d–f**) SL-PRU unmixing. Dashed boxes in (**a**) and (**b**) and in (**c**) and (**d**) indicate the zoom area in **c–d** and **e–f**, respectively. Scale bars equal 100 μm (**d**), 25 μm (**e**), and 10 μm (**f**). Eight taxon-specific probes were used, labeled with the following fluorophores: Alexa Fluor 488 (*Actinomyces* sp.), Alexa Fluor 594 (*Streptococcus* sp.), Alexa Fluor 555 (*Fusobacterium* sp.), Alexa Fluor 660 (*Leptotrichia* sp.), tetrachlorofluorescein (*Neisseriaceae*), Alexa Fluor 647 (*Porphyromonas* sp.), Alexa Fluor 514 (*Veillonella* sp.), Alexa Fluor 633 (*Prevotella* sp.), and Rhodamine Red-X (*Pasteurellaceae*). (Reproduced from Wang et al. [32])

multimodal approaches could bridge the gap between spectral precision and biological interpretability in fluorescence imaging.

1.4 Visualizing Biofilm Architecture with CLASI-FISH

By employing confocal microscopy with CLASI-FISH probes targeting *Fusobacterium* and other major taxa, researchers can achieve high-systems-level visualization of dental plaque biofilm structure (as seen in Fig. 1), facilitating improved data interpretation and the generation of new hypotheses for future studies. CLASI-FISH is a complementary technique for studying *Fusobacterium* in complex oral biofilms, as it enables researchers not only to measure its relative abundance but also to examine its spatial location within biofilms. By simultaneously visualizing *Fusobacterium* and other taxa, the spatial organization driven by *Fusobacterium*'s elongated morphology and its capacity to bind other bacteria through adhesin proteins can be observed.

This chapter outlines a method for visualizing oral biofilms containing *Fusobacterium* species using CLASI-FISH, detailed in subsequent sections. The protocol focuses on sample preparation and image acquisition for dental plaque samples fixed on slides. Researchers may need to adapt this protocol to suit specific research questions, ensuring optimal results. Notably, combining multiple probes is not always necessary when studying a small number of taxa, and the same protocol can be applied regardless of the number of probes used.

The protocol described here, adaptable to dental plaque samples on slides, facilitates investigations into *Fusobacterium*'s interactions with other microorganisms under various conditions, such as exposure to antimicrobial agents or environmental shifts. These capabilities make CLASI-FISH an essential tool for researchers investigating the structural role of *Fusobacterium* in oral microbial ecosystems.

2 Materials

2.1 Solutions and Buffers (see Notes 1, 2, 5, and 7)

1. Phosphate-buffered saline (PBS), pH 7.4.
2. 2% (v/v) paraformaldehyde in water.
3. 100% ethyl alcohol.
4. Hybridization buffer: 0.9 M NaCl, 0.02 M Tris-HCl (pH 7.5), 0.01% (m/v) sodium dodecyl sulfate (SDS), 20% (v/v) HiDi formamide, and 2 μM of each probe.
5. Wash Solution 1: 0.9 M NaCl, 0.02 M Tris-HCl (pH 7.5), 0.01% (m/v) SDS, and 20% (v/v) HiDi formamide (identical to the hybridization buffer except for the absence of probes).
6. Wash solution 2: 0.9 M NaCl, 0.02 M Tris-HCl (pH 7.5), and 0.01% SDS (m/v) (the same as Wash Solution 1 but without formamide).
7. Resuspension buffer: 0.02 M Tris-HCl (pH 7.5).

2.2 Slide Preparation and Cell Labeling

1. Adhesive glass slides.
2. No. 1.5 coverslips.
3. Humid chamber (a pipette tip box with a moist paper towel at the bottom).
4. Hybridization chamber (a 50-mL conical tube lined with a moist paper towel).
5. Antifade mountant (e.g., ProLong Gold).
6. Fluorescent 16S rRNA-binding oligonucleotide probes specific to *Fusobacterium* and other oral biofilm taxa of interest or that universally bind to bacteria, fungi, or archaea for visualization of untargeted taxa (examples can be found in Table 1).

Table 1
Selected 16S rRNA-binding DNA probes for labeling bacteria in oral biofilms

Probe name [Ref]	Target taxon	Probe Sequence (5′-3′)
EUB-338 [34]	Eubacteria (universal)	GCTGCCTCCCGTAGGAGT
FUS-714 [11]	*Fusobacterium* sp.	GGCTTCCCCATCGGGCATT
STR-405 [35]	*Streptococcus* sp.	TAGCCGTCCCTTCTGGT
PRV-392 [36]	*Prevotella* sp.	GCACGCTACTTGGCTGG
PGI-350 [11]	*Porphyromonas* sp.	CCTCACGCCTTACGACGG
COR-633 [22]	*Corynebacterium* sp.	AGTTATGCCCGTTATCGCCTG
ROT-491 [11]	*Rothia* sp.	TAGCCGGCGCTTCTCTG
ACT-476 [37]	*Actinomyces* sp.	ATCCAGCTACCGTCACACC
LEP-568 [11]	*Leptotrichia* sp.	GCCTAGATGCCCTTATG
VEI-488 [38]	*Veillonella* sp.	CCGTGGCTTCTATCCCG
NEI-1030 [11]	*Neisseriaceae*	CCTGGTTACGGCTCCCG
CAP-371 [39]	*Capnocytophaga* sp.	TCAGTCTTCCGACCATTG
PAS-111 [11]	*Pasteurellaceae*	TCCCAAGCATTACTCACC
GEM-572 [11]	*Gemella* sp.	TAAACCACTCGGCGCGCTT
SEL-60 [11]	*Selenomonas* sp.	TCATTGGCTCCGTTCGAC
CAM-1021 [11]	*Campylobacter* sp.	TTTCTGCAGACACACTC
TRP-684 [11]	*Treponema* sp.	TCTACAGATTCCACCCCTAC

2.3 Imaging and Analysis

1. Slides containing *E. coli* strain K12 (or other) grown to mid-log phase and labeled with probes conjugated to individual fluorophores (one slide per fluorophore) for use as a reference in CLASI-FISH image acquisition and spectral unmixing.
2. Spectral confocal microscope.
3. Spectral unmixing software (integrated with the confocal microscope software, if available).
4. ImageJ or equivalent image processing software.

3 Methods

3.1 Slide Preparation and Sample Fixation

1. Rinse adhesive glass slides thoroughly with sterile water and allow them to air-dry.
2. Gently place the oral biofilm sample onto the adhesive slide.

3. Fix the cells in the sample by adding 35 mL of 2% (v/v) paraformaldehyde in MilliQ water (*see* **Notes 1** through **5**) to a 50-mL conical tube, and gently place the slide with the plaque sample into the tube, ensuring the plaque sample is fully submerged.
4. Incubate at room temperature for 1.5 h.
5. Gently replace the 2% (v/v) paraformaldehyde solution with 50 mL MilliQ water.
6. Incubate for 10 min at room temperature.
7. Gently rinse the slide three times with MilliQ water to remove any residual paraformaldehyde, and leave the slide in 50 mL of MilliQ water.
8. Incubate for 10 min at room temperature. Repeat this step twice for three washes.

3.2 Sample Labeling and Mounting

1. Add 100 μL of hybridization buffer containing CLASI-FISH oligonucleotide probes specific to *Fusobacterium* and other taxa of interest (2 μM each) directly on top of the fixed plaque sample.
2. Place a coverslip over the sample to prevent evaporation.
3. Place the slide in a hybridization chamber, and seal it tightly to prevent evaporation of the solution.
4. Incubate at 46 °C for 12–16 h in a hybridization oven, protecting the sample from light by wrapping the hybridization chamber in aluminum foil (*see* **Notes 6** and **7**).
5. In the fume hood, retrieve the slide from the hybridization chamber, and gently remove the coverslip by dipping the slide in a 50-mL conical tube containing Wash Solution 1, allowing it to slip off.
6. Place the slide in a 50-mL conical tube with Wash Solution 1, and incubate at 48 °C for 15 min (*see* **Note 8**).
7. Transfer the slide to a 50-mL conical tube with Wash Solution 2, and incubate at 48 °C for 15 min.
8. Transfer the slide to a 50-mL conical tube with room temperature Resuspension Solution, and incubate at room temperature for 15 min.
9. Dehydrate the slide using an ethanol series. Sequentially incubate the slide in 50-mL conical tubes containing 50% (v/v), 75% (v/v), and twice in 100% (v/v) ethanol, for 10 s each.
10. Allow the slide to air-dry in a dark location.
11. Apply a droplet of antifade mountant and carefully place a No. 1.5 coverslip on top. Take care to avoid the formation of bubbles between the coverslip and the slide.

3.3 Imaging and Spectral Unmixing

1. Prepare reference samples for each fluorophore by fixing *E. coli* and then labeling with the EUB-338 probe (Table 1) conjugated to each fluorophore used to label your plaque specimen as described above (*see* **Note 9**).
2. Acquire spectral images using a laser scanning confocal microscope to capture the full emission spectra emitted from the sample. Use appropriate laser lines, and collect emissions over a broad spectral range to ensure accurate representation of each fluorophore.
3. Image each reference *E. coli* sample using identical microscope settings used for the experimental images (e.g., laser wavelength and power, gain, pinhole, and detection range) to ensure consistency between reference and experimental data.
4. Apply spectral unmixing algorithms using the emission spectra extracted from the reference samples to deconvolve overlapping fluorescence signals in the experimental images. Process the unmixed images to enhance resolution and enable accurate identification of each fluorophore (*see* **Note 10**).

4 Notes

1. Prepare and use solutions containing formamide or paraformaldehyde in a chemical fume hood, as these chemicals are volatile and toxic.
2. Use freshly prepared paraformaldehyde solutions to ensure optimal fixation and probe penetration, and do not exceed the recommended fixation time of 1.5 h, as prolonged fixation may reduce cell permeability to CLASI-FISH probes, particularly in gram-positive cells due to their thick peptidoglycan layers.
3. Dispose of solutions containing formamide or paraformaldehyde in accordance with the appropriate chemical hazard protocols.
4. For labeling tissue samples, formalin fixation is usually preferred; however, extended fixation times needed to preserve tissue architecture may result in over-fixation of bacteria and sub-optimal FISH labeling.
5. Use Milli-Q ultrapure water for preparing all solutions and buffers.
6. During all steps involving fluorophores, care should be taken to minimize the sample's exposure to light, especially sunlight from lab windows.
7. For complex dental plaque samples, note that labeling efficiency may vary between gram-positive and gram-negative

cells due to differences in cell wall permeability. The extended hybridization time (12–16 h) in this protocol optimizes probe penetration for a diverse range of oral bacteria.

8. Pre-warming wash solutions at 48 °C before usage ensures that the samples spend the entire required time at the appropriate temperature during washes.
9. Ensure accurate spectral calibration to minimize crosstalk between fluorophores during unmixing.
10. No single CLASI-FISH protocol is optimal for all organisms. Image analysis requires thresholding to identify taxa based on empirical labeling of cultured isolates.

Acknowledgments

This work was supported by the National Institutes of Health (grant numbers R01DE031213, R01DE030927, S10OD028600).

Conflict of Interest Statement The authors have no conflicts of interest to declare that are relevant to the content of this chapter.

References

1. Robertson LA (2023) 1723-2023: a review to commemorate van Leeuwenhoek's death, 300 years ago. Antonie Van Leeuwenhoek 116(10):919–935. https://doi.org/10.1007/s10482-023-01859-4
2. Wollman JM, Nudd R, Hedlund EG, Leake MC (2015) From animaculum to single molecules: 300 years of the light microscope. Open Biol 5(4):150019. https://doi.org/10.1098/rsob.150019
3. Sarfatis A, Wang Y, Twumasi-Ankrah N, Moffitt JR (2025) Highly multiplexed spatial transcriptomics in bacteria. Science 387(6732): eadr0932. https://doi.org/10.1126/science.adr0932
4. Barbosa A, Miranda S, Azevedo NF, Cerqueira L, Azevedo AS (2023) Imaging biofilms using fluorescence in situ hybridization: seeing is believing. Front Cell Infect Microbiol 13:1195803. https://doi.org/10.3389/fcimb.2023.1195803
5. Amann R, Fuchs BM (2008) Single-cell identification in microbial communities by improved fluorescence in situ hybridization techniques. Nat Rev Microbiol 6(5):339–348. https://doi.org/10.1038/nrmicro1888
6. Morgan XC, Segata N, Huttenhower C (2013) Biodiversity and functional genomics in the human microbiome. Trends Genet 29(1): 51–58. https://doi.org/10.1016/j.tig.2012.09.005
7. Amann RI, Ludwig W, Schleifer K (1995) Phylogenetic identification and in situ detection of individual microbial cells without cultivation. Microbiol Rev 59(1):143–169. https://doi.org/10.1128/mr.59.1.143-169.1995
8. Chen T, Yu W, Izard J, Baranova OV, Lakshmanan A, Dewhirst FE (2010) The human oral microbiome database: a web accessible resource for investigating oral microbe taxonomic and genomic information. Database 2010. https://doi.org/10.1093/database/baq013
9. Wright ES, Yilmaz LS, Corcoran AM, Ökten HE, Noguera DR (2014) Automated design of probes for rRNA-targeted fluorescence in situ hybridization reveals the advantages of using dual probes for accurate identification. Appl Environ Microbiol 80(16):5124–5133. https://doi.org/10.1128/AEM.01685-14
10. Garini Y, Young IT, McNamara G (2006) Spectral imaging: principles and applications.

Cytometry A 69(8):735–747. https://doi.org/10.1002/cyto.a.20311

11. Valm AM, Welch JLM, Rieken CW, Hasegawa Y, Sogin ML, Oldenbourg R, Dewhirst FE, Borisy GG (2011) Systems-level analysis of microbial community organization through combinatorial labeling and spectral imaging. Proc Nat Acad Sci 108(10):4152–4157. https://doi.org/10.1073/pnas.1101134108
12. Valm AM, Welch JLM, Borisy GG (2012) CLASI-FISH: principles of combinatorial labeling and spectral imaging. Syst Appl Microbiol 35(8):496–502. https://doi.org/10.1016/j.syapm.2012.03.004
13. Sakanaka A, Kuboniwa M, Shimma S, Alghamdi SA, Mayumi S, Lamont RJ, Fukusaki E, Amano A (2022) *Fusobacterium nucleatum* metabolically integrates commensals and pathogens in oral biofilms. mSystems 7(4):e00170-22. https://doi.org/10.1128/msystems.00170-22
14. Abusleme L, Hoare A, Hong BY, Diaz PI (2021) Microbial signatures of health, gingivitis, and periodontitis. Periodontol 2000 86(1):57–78. https://doi.org/10.1111/prd.12362
15. Kolenbrander PE, Palmer RJ Jr, Periasamy S, Jakubovics NS (2010) Oral multispecies biofilm development and the key role of cell-cell distance. Nat Rev Microbiol 8(7):471–480. https://doi.org/10.1038/nrmicro2381
16. Shokeen B, Park J, Duong E, Rambhia S, Paul M, Weinberg A, Shi W, Lux R (2020) Role of FAD-I in fusobacterial interspecies interaction and biofilm formation. Microorganisms 8(1):70. https://doi.org/10.3390/microorganisms8010070
17. Guo L, Shokeen B, He X, Shi W, Lux R (2017) *Streptococcus mutans* SpaP binds to RadD of *Fusobacterium nucleatum* ssp. *polymorphum*. Mol Oral Microbiol 32:355–364. https://doi.org/10.1111/omi.12177
18. Kaplan CW, Lux R, Haake SK, Shi W (2009) The *Fusobacterium nucleatum* outer membrane protein RadD is an arginine-inhibitable adhesin required for inter-species adherence and the structured architecture of multispecies biofilm. Mol Microbiol 71(1):35–47. https://doi.org/10.1111/j.1365-2958.2008.06503.x
19. Coppenhagen-Glazer S, Sol A, Abed J, Naor R, Zhang X, Han YW, Bachrach G (2015) Fap2 of *Fusobacterium nucleatum* is a galactose-inhibitable adhesin involved in coaggregation, cell adhesion, and preterm birth. Infect Immun 83(3):1104–1113. https://doi.org/10.1128/iai.02838-14
20. Kinder SA, Holt SC (1993) Localization of the *Fusobacterium nucleatum* T18 adhesin activity mediating coaggregation with *Porphyromonas gingivalis* T22. J Bacteriol 175(3):840–850. https://doi.org/10.1128/jb.175.3.840-850.1993
21. Lima BP, Shi W, Lux R (2017) Identification and characterization of a novel *Fusobacterium nucleatum* adhesin involved in physical interaction and biofilm formation with *Streptococcus gordonii*. MicrobiologyOpen 6(3):e00444. https://doi.org/10.1002/mbo3.444
22. Welch JLM, Rossetti BJ, Rieken CW, Dewhirst FE, Borisy GG (2016) Biogeography of a human oral microbiome at the micron scale. Proc Nat Acad Sci 113(6):E791–E800. https://doi.org/10.1073/pnas.1522149113
23. Neher RA, Mitkovski M, Kirchhoff F, Neher E, Theis FJ, Zeug A (2009) Blind source separation techniques for the decomposition of multiply labeled fluorescence images. Biophys J 96(9):3791–3800. https://doi.org/10.1016/j.bpj.2008.10.068
24. Bhatt JS, Joshi MV (2020) Deep learning in hyperspectral unmixing: a review. In: IGARSS 2020–2020 IEEE international geoscience and remote sensing symposium, pp 2189–2192. https://doi.org/10.1109/IGARSS39084.2020.9324546
25. Bioucas-Dias JM, Plaza A, Dobigeon N, Parente M, Du Q, Gader P, Chanussot J (2012) Hyperspectral unmixing overview: geometrical, statistical, and sparse regression-based approaches. IEEE J Sel Top Appl Earth Obs Remote Sens 5(2):354–379. https://doi.org/10.1109/JSTARS.2012.2194696
26. Heylen R, Parente M, Gader P (2014) A review of nonlinear hyperspectral unmixing methods. IEEE J Sel Top Appl Earth Obs Remote Sens 7(6):1844–1868. https://doi.org/10.1109/JSTARS.2014.2320576
27. Bioucas-Dias JM, Figueiredo MAT (2010) Alternating direction algorithms for constrained sparse regression: application to hyperspectral unmixing. In: 2010 2nd workshop on hyperspectral image and signal processing: evolution in remote sensing, pp 1–4. https://doi.org/10.1109/WHISPERS.2010.5594963
28. Giampouras PV, Themelis KE, Rontogiannis AA, Koutroumbas KD (2016) Simultaneously sparse and low-rank abundance matrix estimation for hyperspectral image unmixing. IEEE Trans Geosci Remote Sens 54(8):4775–4789. https://doi.org/10.1109/TGRS.2016.2551327
29. Guo Z, Wittman T, Osher S (2009) L1 unmixing and its application to hyperspectral image

enhancement. In: Algorithms and technologies for multispectral, hyperspectral, and ultraspectral imagery XV, vol 7334, pp 579–587. https://doi.org/10.1117/12.818245

30. Iordache MD, Bioucas-Dias JM, Plaza A (2012) Total variation spatial regularization for sparse hyperspectral unmixing. IEEE Trans Geosci Remote Sens 50(11): 4484–4502. https://doi.org/10.1109/TGRS.2012.2191590
31. Iordache MD, Bioucas-Dias JM, Plaza A (2013) Collaborative sparse regression for hyperspectral unmixing. IEEE Trans Geosci Remote Sens 52(1):341–354. https://doi.org/10.1109/TGRS.2013.2240001
32. Wang R, Lemus AA, Henneberry CM, Ying Y, Feng F, Valm AM (2023) Unmixing biological fluorescence image data with sparse and low-rank Poisson regression. Bioinformatics 39(4):btad159. https://doi.org/10.1093/bioinformatics/btad159
33. Gharbia SE, Shah HN, Edwards KJ (2015) Fusobacterium. In: Bergey's manual of systematics of archaea and bacteria, pp 1–18. https://doi.org/10.1002/9781118960608.gbm00768
34. Amann RI, Binder BJ, Olson RJ, Chisholm SW, Devereux R, Stahl DA (1990) Combination of 16S rRNA-targeted oligonucleotide probes with flow cytometry for analyzing mixed microbial populations. Appl Environ Microbiol 56(6):1919–1925. https://doi.org/10.1128/aem.56.6.1919-1925.1990
35. Paster BJ, Bartoszyk IM, Dewhirst FE (1998) Identification of oral streptococci using PCR-based, reverse-capture, checkerboard hybridization. Methods Cell Sci 20:223–231. https://doi.org/10.1023/A:1009715710555
36. Diaz PI, Chalmers NI, Rickard AH, Kong C, Milburn CL, Palmer RJ Jr, Kolenbrander PE (2006) Molecular characterization of subject-specific oral microflora during initial colonization of enamel. Appl Environ Microbiol 72(4): 2837–2848. https://doi.org/10.1128/AEM.72.4.2837-2848.2006
37. Gmür R, Lüthi-Schaller H (2007) A combined immunofluorescence and fluorescent in situ hybridization assay for single cell analyses of dental plaque microorganisms. J Microbioll Methods 69(2):402–405. https://doi.org/10.1016/j.mimet.2006.12.012
38. Chalmers NI, Palmer RJ Jr, Cisar JO, Kolenbrander PE (2008) Characterization of a *Streptococcus* sp.-*Veillonella* sp. community micromanipulated from dental plaque. J Bacteriol 190(24):8145–8154. https://doi.org/10.1128/jb.00983-08
39. Zijage V, Van Leeuwen MB, Degener JE, Abbas F, Thurnheer T, Gmür R, Harmsen HJM (2010) Oral biofilm architecture on natural teeth. PLoS One 5(2):e9321. https://doi.org/10.1371/journal.pone.0009321

Chapter 2

Isolation and Characterization of *Fusobacterium nucleatum* Subspecies from Oral Clinical Specimens

Madeline Krieger, Christina Borland, and Justin Merritt

Abstract

The identification and quantification of *Fusobacterium nucleatum* subspecies is of paramount importance to researchers investigating fusobacterial composition in clinical specimens of human health and disease. Here, we present two validated methods to identify and/or quantify oral fusobacteria present within clinical specimens. Importantly, both approaches can unambiguously distinguish the four *F. nucleatum* subspecies and *Fusobacterium periodonticum*. The first method is a convenient PCR-based approach used for rapid genotyping of individual fusobacterial colonies and mixed cultures. The second is a quantitative, next-generation sequencing-based approach that directly measures the different fusobacteria present within complex clinical specimens. Together, these methods provide a comprehensive toolkit for fusobacterial profiling in clinical studies.

Key words *Fusobacterium nucleatum* subspecies, *Fusobacterium* isolation, Dental abscess, Tumor microbiome

1 Introduction

Fusobacterium nucleatum is a ubiquitous member of the oral microbiota that has been traditionally subdivided into four separate subspecies: *F. nucleatum* subspecies *animalis*, *F. nucleatum* subspecies *nucleatum*, *F. nucleatum* subspecies *polymorphum*, and *F. nucleatum* subspecies *vincentii* [1, 2]. These four subspecies have been historically regarded as physiologically equivalent, and *F. nucleatum* subspecies *nucleatum* has served as the principal representative employed in most laboratory investigations of fusobacterial pathobiology. Recently, multiple studies have challenged this common perception of *F. nucleatum* by demonstrating clear distinctions in the physiology and pathogenic potential of *F. nucleatum* subspecies [3–5]. Of particular note, *F. nucleatum* subspecies *animalis* appears to be the primary organism enriched in many different oral and extraoral diseases, especially those

Hung Ton-That (ed.), *Fusobacterium nucleatum: Methods and Protocols*, Methods in Molecular Biology, vol. 3055, https://doi.org/10.1007/978-1-0716-5388-3_2,

associated with inflammation, such polymicrobial abscesses and colorectal cancer [4–9]. Furthermore, several recent publications have also provided compelling evidence that *F. nucleatum* subspecies are in fact separate species [5, 10]. For these reasons, there is a strong rationale to reevaluate previous assumptions about fusobacterial pathobiology, while current and future clinical studies should specifically distinguish among the fusobacteria detected in clinical specimens. To support such efforts, we present two validated methods to identify and/or quantify oral fusobacteria in clinical studies. The first is a convenient PCR-based approach used for rapid genotyping of individual fusobacterial colonies and mixed cultures. The second is a quantitative, next-generation sequencing-based approach that measures the different fusobacteria present within complex clinical specimens. Together, these methods provide a comprehensive toolkit for fusobacterial profiling in clinical studies.

2 Materials

2.1 Sample Collection and Bacterial Culturing

1. Medium supplement stocks: 5 mg/mL of hemin dissolved in 0.1 M NaOH, 10 mg/mL of vitamin K dissolved in ethanol, 10 mg/mL of N-acetylmuramic acid, laked sheep blood (lysed with three freeze-thaw cycles and then centrifuged to remove cellular debris), fetal bovine serum (FBS) (heat treated at 50 °C for 1 hr.), 50% (w/v) sterile-filtered sucrose in ddH_2O, 10 mg/mL of kanamycin, 50 mg/mL of vancomycin, agar, pancreatic digest of casein, proteose peptone, yeast extract, L-cysteine, L-glutamate, sodium chloride, autoclaved 100% (v/v) glycerol, and oxyrase. Optional: L-cysteine, sodium thioglycolate.
2. Modified SHI (mSHI) medium: To make 500 mL, dissolve 5 g of protease peptone, 2.5 g of bacto-tryptone, 2.5 g of yeast extract, 1.25 g of KCl, 0.03 g of urea, and 0.085 of g arginine in 423 mL of distilled-deionized water with stirring. Combine and autoclave at 121 °C for 15 min. Allow to cool to 50–55 °C in a water bath, and then add 0.5 mL of hemin, 50 μL of vitamin K, 0.5 mL of N-acetylmuramic acid, 25 mL of laked sheep blood, 50 mL of FBS, and 1 mL sucrose. Thoroughly mix and store at 4°C.
3. mSHI transfer medium: Make a 500-mL modified SHI medium as outlined above, with the addition of 0.375 g of agar prior to autoclaving (increases the viscosity of the medium and protects from oxygen infiltration after sampling) (*see* **Note 1**).
4. MCDC medium: To make 500 mL, combine 8 g of agar, 5 g of pancreatic digest of casein, 5 g of proteose peptone, 2.5 g of yeast extract, and 2.5 g of NaCl in 474-mL ddH_2O with stirring. Combine and autoclave at 121°C for 15 min. Allow

to cool to 50–55 °C in a water bath, and then add 200 mg of L-cysteine, 1 g of L-glutamate, 25 mL of laked sheep blood, 0.5 mL of vitamin K, and 0.5 mL of hemin. To specifically select for fusobacteria, add 0.5 mL of kanamycin and 75 of μL vancomycin.

5. Disposable sterile inoculating loops (1 μL).
6. Sterile petri dishes (100 mm × 15 mm).
7. Sterile polyester tipped swabs.
8. 1.5-mL polypropylene tubes.
9. Anaerobic chamber (85% N_2, 10% CO_2, 5% H_2) at 37 °C.
10. Cryovials.
11. Vortex mixer.

2.2 Kit-Based DNA Extraction Supplies

1. ZymoBIOMICS DNA isolation kit (Zymo).
2. Bead beating mixer (e.g., Bertin Precellys Evolution Touch Homogenizer).
3. Microcentrifuge (e.g., Eppendorf model 5418 or equivalent).

2.3 Alternative DNA Extraction Supplies

1. DNA-extraction reagents: phenol:chloroform:isoamyl alcohol (25:24:1), STES buffer with lysozyme (0.5 M of NaCl, 0.02 M of EDTA, 0.2 M of Tris-HCl (pH 8), 20 mg/mL of lysozyme), 1% (w/v) SDS, 0.1 mg/mL of proteinase K, 50 μg/mL of RNase A, 3 M of sodium acetate (pH 5.5), and 100% (v/v) and 70% (v/v) ethanol.
2. Water bath at 55°C.
3. Microcentrifuge (e.g., Eppendorf model 5418 or equivalent).

2.4 PCR-Based Subtyping

1. **PCR supplies**: molecular biology grade double distilled water (ddH_2O), high fidelity polymerase (e.g., Phusion and corresponding buffers, Fisher Scientific), 10 mM of dNTP stock, template DNA, forward and reverse primers, thermocycler, and compatible PCR tubes.
2. **Agarose gel electrophoresis supplies**: TAE buffer (40 mM of Tris-HCl base, 20 mM of acetic acid, 1 mM of EDTA), 1.0% (w/v) agarose in TAE buffer, ethidium bromide for visualization, 1 kb of DNA standard ladder, agarose electrophoresis chamber systems, and appropriate power supply.

2.5 Next-Generation Sequencing Supplies

1. AMPure XP beads (Beckman Coulter).
2. 96-well magnetic plate (e.g., CleanMag magnetic plate or equivalent).
3. Freshly prepared 70% ethanol.

3 Methods

3.1 Sample Collection

1. The following method describes sample collection from oral biofilm and abscess specimens. Adjust the methodology as appropriate for other sample types.
2. Assemble cryovial collection tubes with approximately 1 mL of mSHI transfer media. Pre-reduce both media and cryovials in an anaerobic chamber for at least 24 hr. prior to use.
3. Seal cryovial firmly, and wrap with parafilm before removing from the anaerobic chamber.
4. Remove vials from the anaerobic chamber within several hours of intended usage. In our experience, vials may stay sufficiently reduced for up to 5 days in ambient air, but oxygen infiltration can introduce variability in the results. Store tubes at room temperature.
5. Uncap the tube immediately before clinical sample collection. Using a sterile cotton swab, collect abscess or dental plaque material (*see* Fig. 1a) (*see* **Note 2**). Break off or cut the sterile swab tip into the cryovial and reseal the tube (*see* Fig. 1b).
6. Vortex the cryovial to resuspend swab contents in the tube (either before or after returning it to the anaerobic chamber) (*see* Fig. 1c).
7. Return cryovial to the anaerobic chamber as quickly as possible (*see* Fig. 1d). In our experience, several hours of delay did not cause a noticeable loss in viable *F. nucleatum* recovery.
8. For *Fusobacterium* isolation: proceed to bacterial culturing (*see* Subheading 3.2).
9. Inside the chamber, transfer the resuspended sample to a new cryovial. The swab will have absorbed some of the medium, so you can expect to collect slightly less medium than original aliquoted.
10. Dilute to 25% (v/v) with autoclaved glycerol in H_2O. Mix thoroughly.
11. Store at −80°C (*see* Fig. 1e).

3.2 Bacterial Culture

1. If directly using a freshly collected specimen, spread 100 μL of resuspended sample to thoroughly coat the plate. If isolating from a glycerol stock, add sufficient sample to thoroughly coat the plate (*see* Fig. 1e).
2. Incubated at 37°C for 3–4 days in an anaerobic chamber.
3. Collect selected colonies or all of the microbial growth on the plate with a sterile loop (*see* Fig. 2a).
4. Resuspend the cells in sterile PBS to an optical density OD_{600} of 1.0.

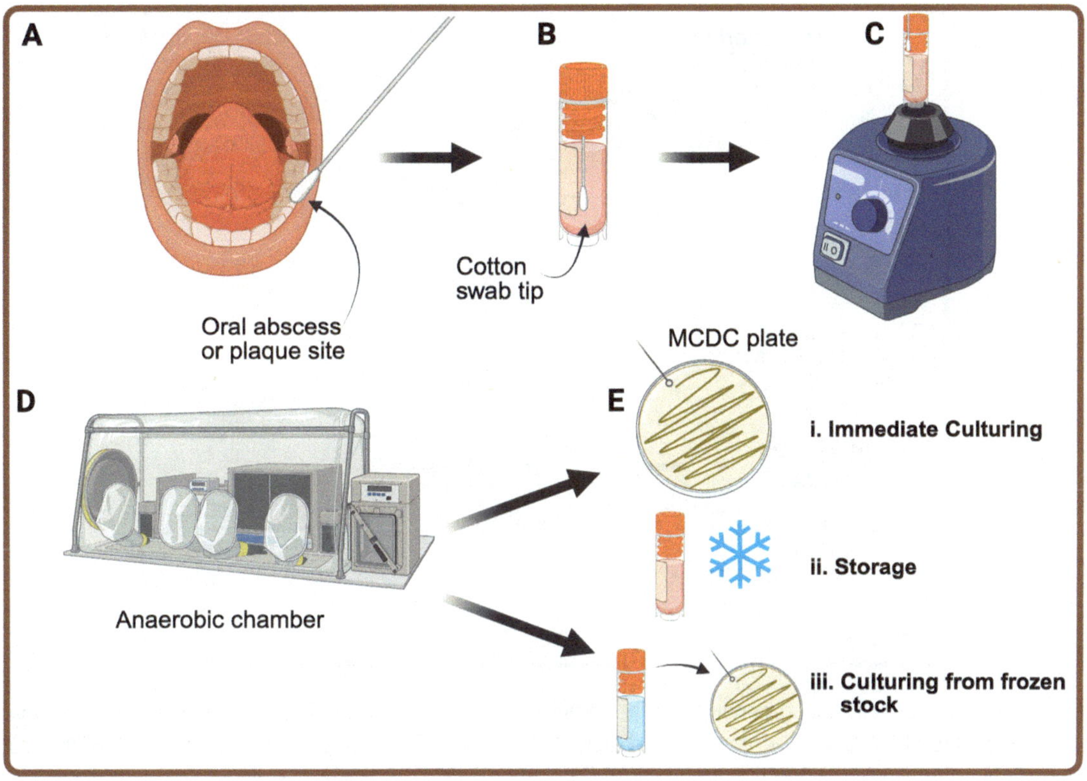

Fig. 1 Overview of clinical sample collection procedure. (**a**) Transfer patient sample directly into pre-reduced mSHI transfer media. (**b**) Break off or cut the sterile swab tip into the cryovial and securely cap to minimize oxygen exposure. (**c**) Vortex sample thoroughly to resuspend the clinical material in the vial. (**d**) Move the tube to an anaerobic chamber. (**e**.i) For immediate culturing, spread ~100 μL of resuspended sample onto a pre-reduced MCDC plate. Incubate in anaerobic chamber at 37°C for 3–4 days, monitoring for colony formation. (**e**.ii) To store a sample, transfer the resuspended sample to a new cryovial. Add 25% (v/v) glycerol, mix, and store at −80°C. (**e**.iii) To culture from a frozen stock, transfer the cryovial from a −80°C freezer into the anaerobic chamber using an appropriately sized aluminum heat block insert that has been precooled to −80°C. Uncap the vial and then use a sterile inoculating loop to carefully scoop ~50–100 μL of frozen glycerol stock. Spread the inoculum evenly across a pre-reduced MCDC plate, and then incubate in an anaerobic chamber at 37°C for 3–4 days

3.3 DNA Extraction (Kit-Based)

1. We recommend using the ZymoBIOMICS DNA Miniprep kit with bead beating per the manufacturer's instructions, using a double-elution protocol with 50–100 μL of H_2O prewarmed to 60 °C without a Zymo-Spin III-HRC filtering step (*see* Fig. 2b).

3.4 DNA Extraction (Alternative)

An alternative to the kit-based protocol adapted from Cho et al. [11] is provided below.

1. Centrifuge cells at 4000 × g for 10 min. to pellet. Remove supernatant.
2. Resuspend in STES buffer with lysozyme.
3. Incubate in a water bath at 55°C for 3 hr.

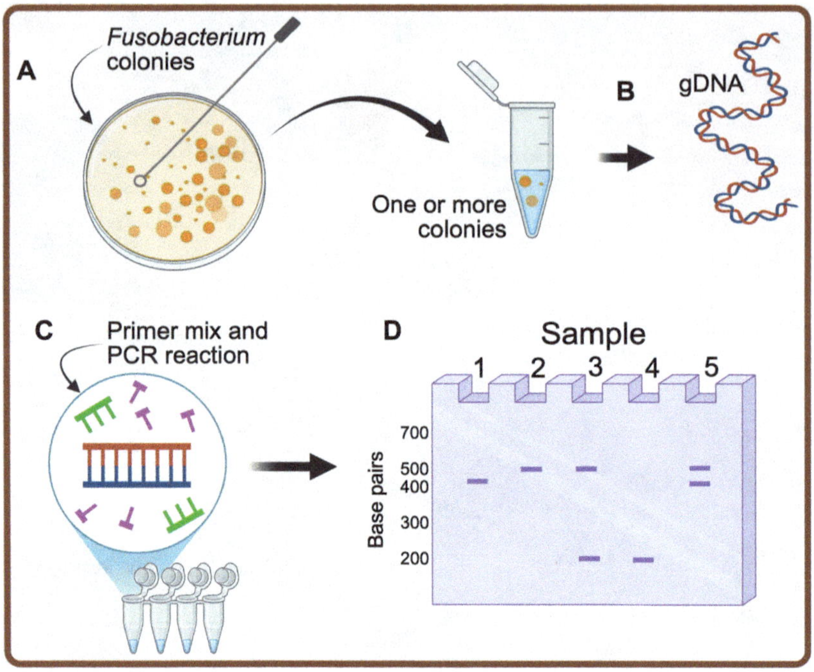

Fig. 2 Overview of PCR-based genotyping. (**a**) Select one or more colonies from an inoculated MCDC agar plate incubated at 37°C for 3–4 days in anaerobic conditions. (**b**) Extract gDNA using either a kit-based or phenol/chloroform-based extraction protocol. (**c**) Combine gDNA, PCR mastermix, and primers depending on subspecies detection goals. Separate PCR amplicons using agarose gel electrophoresis. (**d**) Compare product size(s) to known *F. nucleatum* size standards to determine subspecies genotype(s) within the sample

4. Add SDS to a 1% (w/v) final concentration followed by proteinase K (final concentration of 0.1 mg/mL), and mix well.
5. Incubate at 55°C for 1–2 hr.
6. Add 1 volume of 25 phenol: 24 chloroform:1 isoamyl alcohol (pH 6.4). Mix gently.
7. Spin for 5 min. at 16,000 × g.
8. Collect aqueous layer (top). Dispose of organic layer.
9. Add RNase A to a final concentration of 50 μg/mL to the collected aqueous layer and incubate at 37°C for 2 hr.
10. Add 1 volume of phenol:chloroform:isoamyl alcohol. Mix gently.
11. Centrifuge for 5 min. at 16,000 × g.
12. Collect aqueous layer (top). Dispose of organic layer.
13. Add 1/10 volume of 3 M of sodium acetate (pH 5.5) to aqueous layer.
14. Add three volumes of 100% ice-cold ethanol.
15. Incubate at −20°C for ≥24 hr.
16. Centrifuge at 16,000 × g at 4°C for 30 min.

17. Wash pellet with ice-cold 70% ethanol.
18. Centrifuge at 16,000 × g at 4°C for 15 min.
19. Dry DNA pellet in a laminar flow hood.
20. Resuspend in buffer of choice.

3.5 PCR-Based Genotyping

PCR-based genotyping can identify fusobacterial composition using separate subspecies-specific primer pairs or in a multiplex cocktail to genotype multiple subspecies simultaneously.

1. Prepare the PCR mastermix (*see* Table 1) using the appropriate primers (*see* Table 2). To multiplex and detect multiple subspecies in the same reaction, add each PCR primer to the mixture to reach a final concentration of 0.5 μM in a final reaction volume of 50 μL per 1× reaction (*see* Fig. 2c).
2. Run the following thermocycler protocol: 1 cycle of 95°C for 2 min.; 30 cycles of 95°C for 30 s, 55°C for 30 s, and 72°C for 45 s; and a final extension at 72°C for 5 min.
3. Analyze the PCR products by gel electrophoresis, and compare the product sizes to genotype *Fusobacterium* subspecies present within the specimen. The use of a positive control reaction containing verified fusobacterial DNA from representative subspecies may be helpful to validate the reaction results (*see* Fig. 2d).

3.6 Amplicon Preparation for Next-Generation Sequencing

1. Isolate genomic DNA as described above.
2. Using the primers designed to amplify an 881 bp segment of the zinc protease gene (*znpA*) (*see* Table 3), mix the first round of PCR reactions as outlined in Table 4. We recommend using 4 μL of template DNA in this first reaction, but this volume may be adjusted depending on the concentration of isolated gDNA (*see* Note 3).

Table 1
PCR mastermix recipe for PCR-based genotyping

Reagent	μL for 1× reaction
GoTaq DNA polymerase (Promega)	0.25
10 μM of forward and reverse primers	2.5
25 mM of $MgCl_2$	4
5× reaction buffer	10
10 mM of dNTP mixture	1
DNA	4
H_2O	Adjust to 50 μL

Table 2
Subspecies specific primers for PCR-based genotyping

Target	Specificity	Primer 1 (5′-3′)	Primer 2 (5′-3′)	Product size (pb)
Hypothetical protein	*Fn. animalis*	TTTGTATTTGACAA TTTAATTAGAGCA	TGGGATAATAATCCAAA TATYGCAC	413
Ankyrin repeat proteins	*Fn. nucleatum*	CTTCATTGTTAGGA TTACAAGAA TAGGC	TTTCAGTAACATTTA TTGGGAGGTTT	497
YARHG domain-containing protein	*F. periodonticum*	TTCTATAGTTACC TTCAGCAACTAAA TTAC	AAGAATAGAGAATATTA TACTAATGGAAAC TTAAA	200
Hypothetical protein	*Fn. polymorphum*	AAGATTAAATAAAGA TGGAAATAACCA TRAA	CCATCTCCAAAGCC TTTCC	807
Hypothetical protein	*Fn. vincentii*	CTTTGTCTTATGA TTTATTGCRTATC TTC	GAAGAATCAGAAGCAC TACCAGCA	303

3. Perform amplification using the following thermocycler protocol: 1 cycle of 95°C for 2 min.; 28 cycles of 95°C for 15 s, 55°C for 30 s, and 72°C for 45 s; and a final extension at 72°C for 2 min.
4. Clean amplicons with AMPure XP beads (Beckman Coulter):
   (a) Move 40 μL of each well into a PCR plate.
   (b) Resuspend AMPure XP beads thoroughly.
   (c) Add 72 μL of beads per well (1.8× reaction volume), and pipette up and down (~10 times) to mix thoroughly.
   (d) Incubate mixed samples at room temperature for 5 min.
   (e) Place reaction plate onto a magnetic separator and incubate for 2 min.
   (f) While leaving the plate on the magnet, aspirate and discard the supernatant, leaving 5 μL of supernatant behind.
   (g) With the plate still on the magnet, add 200 μL of 70% ethanol (mixed fresh), and incubate for 30 s at room temperature. Remove ethanol.
   (h) Repeat ethanol wash and remove supernatant. Dry at room temperature for 10 min.
   (i) Remove reaction plate from the magnet, add 40 μL of H_2O to the beads, and pipette up and down (~10 times) to mix. Incubate for 5 min.
   (j) Place plate on the magnet for 2 min. Transfer eluate to a new plate.

Table 3
Primers for PCR amplification of zinc protease A gene amplicons

PCR round	Primer 1 (5′ → 3′)	Primer 2 (5′ → 3′)	Product size (bp)
1	GATGAAGAAAGYATAGANAAAGAAAGAAATGT	TCTCCTACTATTGTTTGTGAATAA	881
2	TCGTCGGCAGCGTCAGATGTGTATAAGAGACAGatagaagaaataaraatgtatga	GTCTCGTGGGCTCGGAGATGTGTATAAGAGACAGcctctttrttctcttatrttttg	467

Table 4
PCR mastermix recipe for NGS-based genotyping and *Fusobacterium* quantification

Reagent	μL for 1× reaction
Taq polymerase (Promega)	0.25
10 μM of forward and reverse primers (first round)	2.5
25 mM of $MgCl_2$	4
5× reaction buffer	10
10 mM of dNTPs	1
H_2O and DNA	Adjust to 50 μL

5. Remove 20 μL of purified PCR products and transfer to a new PCR plate for use in a second round of PCR. Prepare 50 μL of PCR reactions using the same mastermix protocol outlined in step 2 above, but with the PCR round 2 primers. Perform PCR with the following four-step thermocycler protocol: 1 cycle of 95°C for 2 min., 10 cycles of 95°C for 15 s, 42°C for 30 s, and 72°C for 45 s; 15 cycles of 95°C for 15 s, 65°C for 30 s, and 72°C for 45 s; and a final extension at 72°C for 2 min. Each reaction will generate a 467 bp zinc protease amplicon with attached i5/i7 indices for sequencing.
6. Clean amplicons with AMPure XP beads using the same protocol from step 4.
7. Perform PCR using the primers listed in PCR Round 2 above, which amplify the inner region of the zinc protease gene and attached Illumina sequencing adapters. Use the same mastermix formulation in step 2 with the following thermocycler protocol: 1 cycle of 95°C for 2 min.; 8 cycles of 95°C for 15 s, 55°C for 30 s, and 72°C for 45 s; and a final extension at 72°C for 2 min.
8. Clean the resulting DNA with AMPure beads as outlined in step 4.
9. Submit cleaned amplicons for Next-generation sequencing (Illumina MiSeq paired-end or similar).

3.7 Analysis of Next-Generation Sequencing Reads

We adapted the use of DADA2/Qiime2 for processing microbiome samples to classify fusobacterial zinc protease amplicons into distinct subspecies.

1. Install the latest release of the Qiime2 amplicon sequencing distribution on conda. Instructions and specific release information can be found at https://docs.qiime2.org/2024.10/install/native/.

2. Activate the Qiime2 conda environment.
3. Create a zinc protease classifier:
   (a) Create a fasta file with *znpA* sequences from each of the four *F. nucleatum* subspecies and *F. periodontium*, and then import. An example file is shown below (*see* **Notes 4** and **5**).

```
>1
GTATAGAAAAAGAAAGAAATGTGATCATAGAAGAAATAAAAATGTATGAAGATATTCCT-
GAAGAAATAGT
ACATGAAAAAAATGTTGAATATGCTTTAAGAGGAGTACATTCAAATTCAATATCTGGTA-
CAGTTGCAAGT
TTAAAAAAGATAAACAGAAAAGCAATTTTAAATTATTTAGAAAAATATTATGTAGCT-
GAAAATTTAGTTA
TTGTTGCATCTGGAAATATAGATGAAAAATATCTATATAAAGAATTAAACAAAAAAAT-
GAAAAATTTTAG
AAAAACTAAAAAAGAAGAAGTGTTAGATTTATCTTATGAAATAAAAAAAGGTAA-
GAAAGTTGTTAAGAAA
CCTTCAAATCAAATACATCTTTGTTTTACTACAAGAGGAGTTTCTTCAAAGTCAGAAC-
TAAGATATCCAG
CAGCAATAATTTCAAATGTTTTAGGTGAAGGAATGAGTTCAAGACTATTTCAAAAAATAA-
GAGAAGAAAG
AGGACTTGCTTATTCAGTTTATACTTATCTTACAAGATTTGAGAATTGTGGAT-
TACTTTCTGTATATGTT
GGAACTACAAAAGAAGACTATAAAGAAGTTATAAAACTTATAAAAGAAGAGTTTAAAAA-
TATTAAAGAAA
ATGGTATATCAGAAAGAGAGCTAAGAAAAGCTAAGAATAAATATGAAAGTGCTTTTA-
CATTTAGTTTAGA
AAGCACAAGTTCAAGAATGAATAGATTAGCCTCTACATATATTATTTATGGAAAAATTA-
TAAGTCTTGAT
AAGGTTAGAGAAGATATAGAAAAAGTAACTTTAAAAGATATTAAAAAAGCTGCA-
GAATTTTTATTTGATG
AACAATTTTATTCACAAACAATAGTAGGAGATATATAA

>2
GTATAGAAAAAGAAAGAAATGTGATTATAGAAGAAATAAAAATGTATGAAGATATTCCA-
GAAGAGATAGT
ACATGAAAAAAATGTTGAGTATGCTTTAAGAGGAATACATTCAAATTCAATATCTGGAA-
CAGTTGCAACT
TTAAAAAAAATAGATAGAAAAGCTATTTTAAACTATTTAGAAAAACATTATGTAGCT-
GAAAATTTAGTTA
TTGTTGCCTCTGGAAATATAGATGAGAAGTATCTATACAAAGAACTAAATAAAAAAAT-
GAAAAACTTTAG
AAAGTCTAAAAAAGAAGAAATATTAGATTTAACTTATGAAATAAAAAAAGGTAA-
GAAAATTGTAAAGAAA
CCTTCAAACCAAATACATCTTTGTTTTACTACAAGAGGAGTTTCTTCAAATTCAAAAT-
TAAGATATTCTG
```

CTGCAATAATTTCAAATATTTTAGGTGAAGGAATGAGCTCAAGATTATTTCAAAAAATAA-
GAGAAGAAAG

AGGTCTTGCTTACTCAGTTTATACTTATCTTACAAGATTTGAAAATTGTGGGC-
TACTCTCTGTGTATGTT

GGAACTACAAAAGAAGATTATAAAGATGTTATAAAACTTATAAAAGAAGAATTTAAAAA-
TATTAAAGAAA

ATGGTATATCAGAAAGAGAATTAAGAAAAGCTAAGAATAAATATGAAAGTGCCTTTA-
CATTTAGTTTAGA

AAGTACAGGTTCAAGAATGAATAGACTAGCTTCAATGTATGTAACTTATGGAAAAATAA-
TAAGTCTTGAT

AAGGTAAGAGAAGATATTGAAAAGGTTACTTTAAAAGATATTAAAAAGGCTGCT-
GAATTCTTATTTGATG

AACAATATTATTCACAAACAATAGTAGGAGATATTTAA

>3

GTATAGAAAAAGAAAGAAATGTGATTATAGAAGAAATAAAAATGTATGATGATATTCCA-
GAAGAAATTGT

ACATGAAAAAAATATTGAATATGCTTTAAGAGGTATACATTCAAACTCTATTTCTGGAA-
CAGTTTCAAGT

TTAAAAAAGATAGATAGAAAAGCAATTTTAAATTATTTAGAAAAACATTATGTGGCA-
GAAAATTTAGTTA

TAGTTGTAGCTGGAAATATAGATGAAAAATATCTATATAAAGAACTAAATAAAAGAAT-
GAAAGATTTTAG

AAAAGCAAAGAAAAAAGAGGTATTAGATTTAACTTATGAAATAAAAAAAGGTAA-
GAAAGTTGTAAAGAAA

CCTTCAAATCAAATACATCTATGCTTTACTACAAAAGGAGTTTCTTCAAAATCTGATT-
TAAGATACCCAG

CTGCAATAATTTCAAATATCTTAGGTGAAGGAATGAGTTCAAGATTATTTCAAAAGATAA-
GAGAAGAAAG

AGGACTTGCTTATTCAGTTTATACCTATCTTACAAGATTTGAAAATTGTGGAC-
TACTTTCTGTATATGTT

GGGACAACAAAAGAAGACTATAACGAAGTTATAAAACTTATAAAAGAAGAGTTTAATAA-
TATTAAAGAAA

ATGGTATATCAGAAAGAGAACTAAGAAAAGCTAAGAATAAATATGAAAGTGCATTTA-
CATTTAGTTTAGA

AAGCACAAGTTCAAGAATGAATAGATTAGCTTCAACATATATTACTTATGGAAAGATCAT-
TAGTCTTGAT

AAAGTAAGAGAAGATATAGAAAAAGTTACTTTAAAAGATATTAAAAAAGCTGCT-
GATTTTTTATTTGATG

AACAATTTTATTCACAAACAATAGTAGGAGATATTTAA

>4

GTATAGAAAAAGAAAGAAATGTGATTATAGAAGAAATAAAAATGTATGAGGACATACCA-
GAGGAAATAGT

GCATGAAAAAAATGTTGAATATGCTTTAAGAGGAATACATTCAAATTCTATTTCTGGAA-
CAGTTGCAAGT

TTAAAAAAGATAGATAGAAAAGCAATTTTAAATTATTTAGAAAAACATTATGTAGCT-
GAAAATTTAGTTA

TAGTTGCTTCTAGAAATATAGATGAAAAATATCTGTATAAAGAACTAAATAAAAAAAT-
GAAAAATTTTAG
AAAAATTAAAAAAGAAGAAATATTAGATTTATCTTATGAAATAAAGAAAGG-
TAAAAAAGTTGTAAAGAAA
CCTTCAAACCAGATACATCTTTGTTTTACTACAAGAGGAGTTTCTTCAAAGTCAGATT-
TAAGATATCCAG
CTGCAATAATTTCAAATGTTTTAGGTGAAGGGATGAGTTCAAGATTATTTCAAAAGATAA-
GAGAAGAAAG
AGGACTGGCTTATTCAGTTTATACTTACCTTACAAGGTTTGAAAATTGTGGAT-
TACTTTCTGTATATGTT
GGAACTACAAAAGAAGATTATAAAGAAGTTATAAAACTTATAAAAGAAGAATTTAAAAA-
TATTAAAGAAA
ATGGTATATCAGAAAGAGAATTGAGAAAAGCTAAAAATAAATATGAAAGTGCTTTTA-
CATTTAGTTTAGA
AAGCACAAGTTCAAGAATGAACAGATTAGCTTCTACATATATTACTTATGGTAAGATTAT-
TAGCCTTGAT
AAAGTGAGAGAAGATATAGAAAAAGTTACTTTAAAAGATATTAAGAAAGCTGCT-
GAATTTTTATTTGATG
AACAATTTTATTCACAAACAATAGTAGGAGATATATAA

>5
GTATAGAAAAAGAAAGAAATGTGATTATAGAAGAAATAAGAATGTATGAAGACATACCT-
GAAGAAATAGT
TCATGAAAAAAATATTGAATTTGCCTTAAAAGGAATACATTCAAACTCTATCTCAGGAA-
CAATAGCAAGT
TTAAAGAAAATTAATAGAAAAGCTATTTTAAAATATTTAGAAGAACATTATGTAGCTGA-
GAACTTAGTTA
TAGTAGTATCTGGAAATATAGATGAAAAATATTTATATAAAGAATTAAGTAAAAAAAT-
GAAAGACTTTAG
AAGAGCAAAGAAAGAAGAAGTTTTAGATTTAACTTATCAAATCAAAAAAGGTAAAAAAG-
TAGTTAAAAAA
CCTTCTAATCAGATACATCTTTGCTTTACAACAAGAGGAGTTTCTAATAAGTCAGAAT-
TAAGATATCCAG
CAGCTATAATTTCAAATATCTTAGGTGAAGGAATGAGCTCAAGATTGTTTCAAAAAATAA-
GAGAAAAAAG
AGGGCTTGCATATTCAGTTTATACTTATCTGACAAGATTTACTAACTGTGGAC-
TACTTTCTGTTTATGTT
GGAACAACAAAAGAAGACTATAAGGAAGTTATAAAACTTATAAAAGAAGAATTTAAGAA-
TATTAAAGAAA
ATGGAATATCAGAAAGAGAACTTAGAAAAGCTAAAAATAAATATGAAAGTGCCTT-
TACTTTTAGCTTAGA
AAGTACAAGTTCAAGAATGAATAGATTAGCTTCAACATATTTAACTTATGGAGAAATAA-
TAAGTCTTGAT
AAGGTGAGAGAAGATATTGAAAAAGTTAGTCTAAAAGATATAAAAAAGGCAGCA-
GAATTTTTATTTGATG
AAGAATATTACTCACAAACAATAGTAGGAGATATATAA

```
qiime tools import \
  --type 'FeatureData[Sequence]' \
  --input-path <classifier_folder>/Fuso_ZincProt.fasta \
  --output-path <classifier_folder>/Fuso_zincprot.qza
```

(b) Create a reference taxonomy text document that matches the identities of the fasta file above, and import. An example is shown below:

```
1. k__Bacteria; p__Fusobacteriota; c__Fusobacteriia; o__Fusobacteriales; f__Fusobacteriaceae; g__Fusobacterium; s__nucleatum subsp. nucleatum
2. k__Bacteria; p__Fusobacteriota; c__Fusobacteriia; o__Fusobacteriales; f__Fusobacteriaceae; g__Fusobacterium; s__nucleatum subsp. polymorphum
3. k__Bacteria; p__Fusobacteriota; c__Fusobacteriia; o__Fusobacteriales; f__Fusobacteriaceae; g__Fusobacterium; s__nucleatum subsp. vincentii
4. k__Bacteria; p__Fusobacteriota; c__Fusobacteriia; o__Fusobacteriales; f__Fusobacteriaceae; g__Fusobacterium; s__nucleatum subsp. animalis
5. k__Bacteria; p__Fusobacteriota; c__Fusobacteriia; o__Fusobacteriales; f__Fusobacteriaceae; g__Fusobacterium; s__periodonticum
```

4. Import the raw NGS reads:

```
qiime tools import \
  --type 'SampleData[PairedEndSequencesWithQuality]' \
  --input-path <rawreads_directory> \
  --input-format CasavaOneEightSingleLanePerSampleDirFmt \
  --output-path demux-paired-end.qza
```

5. Run DADA2 to process reads:

```
qiime dada2 denoise-paired \
  --i-demultiplexed-seqs /demux-paired-end.qza \
  --p-trunc-len-f 265 \
  --p-trunc-len-r 220 \
  --p-trim-left-f 25 \
  --p-trim-left-r 25 \
  --p-n-threads 0 \
  --output-dir <trimmed_reads_directory>  \
  --verbose
```

6. Generate representative sequences:

```
qiime metadata tabulate \
--m-input-file <trimmed_reads_directory>/representative_sequences.qza \
--o-visualization <trimmed_reads_directory>/representative_sequences.qza
```

7. Add metadata:

```
qiime feature-table summarize \
  --i-table <trimmed_reads_directory>/table.qza \
  --o-visualization <trimmed_reads_directory>/table.qzv \
  --m-sample-metadata-file metadata.txt
```

8. Classify reads and add taxonomy to the representative sequences (*see* **Note 6**).

```
qiime feature-classifier classify-consensus-blast \
  --i-query <trimmed_reads_directory>/representative_sequences.qza \
  --i-reference-reads <classifier_folder>/Fuso_zincprot.qza \
  --i-reference-taxonomy <classifier_folder>/Fuso_ZincProt_ref-taxonomy.qza \
  --p-perc-identity 0.7 \
  --p-maxaccepts 1 \
  --o-classification <trimmed_reads_directory>/taxonomy.qza \
  --verbose


qiime metadata tabulate \
  --m-input-file <trimmed_reads_directory>/representative_sequences.qza \
  --m-input-file <trimmed_reads_directory>/taxonomy.qza \
  --o-visualization Qiime2/Trimmed/taxonomy.qzv
```

9. Prepare biom table for import into R and subsequent analysis with Phyloseq [12].

```
biom add-metadata \
  -i <trimmed_reads_directory>/feature-table.biom \
  -m metadata.tsv \
  --observation-metadata-fp <trimmed_reads_directory>/taxonomy.tsv \
  -o <trimmed_reads_directory>/feature_table_w_taxonomy.biom \
  --sc-separated taxonomy \
  --observation-header OTUID,taxonomy,confidence


biom convert \
-i <trimmed_reads_directory>/feature_table_w_taxonomy.biom \
-o <trimmed_reads_directory>/feature_table_w_taxonomy.tsv \
--to-tsv \
--header-key taxonomy
```

10. Analyze reads using Phyloseq. The user may prefer custom plots. Shown below is a general workflow for importing information from Qiime2 and constructing a basic phyloseq artifact.

```
ps1 = "<trimmed_reads_directory>/feature_table_w_taxonomy.biom"

myData_ps1 = import_biom(ps1)


metadata = import_qiime_sample_data("metadata.txt")

colnames(tax_table(myData_ps1)) = c("Kingdom", "Phylum", "Class", "Order",

"Family", "Genus", "Species")


phylo_obj = merge_phyloseq(myData_ps1,metadata)


#Create a histogram of sample sums

hist(sample_sums(phylo_obj), main="Histogram: Read Counts", xlab="Total Reads",

    border="blue", col="green", las=1, breaks=12)


#Remove any samples with less than the user defined number of reads.

phylo_obj_f_pruned <-

prune_samples(sample_sums(phylo_obj_f)>=<user_defined_threshold>, phylo_obj_f)
```

4 Notes

1. If *Fusobacterium* recovery is low, mSHI may be supplemented with 0.5 g of L-cysteine per 500 mL. Cryovials with mSHI used for sample collection may also be supplemented with 50 μL of oxyrase and 10 μL of 50-mg/mL sodium thioglycolate to further protect against medium oxidation and to improve yields.
2. In the event of a draining abscess, purulence was sampled from the site of the fistula.
3. With the increased publication of fusobacterial genomes, we encourage the user to confirm the binding of these primer sequences to the target genomes. If needed, primer sequences can be modified to ensure subspecies-specific primer binding.
4. Our protocol uses the qiime feature-classifier classify-consensus-blast method to classify reads. However, the user can also utilize the naive bayes classifier method as an alternative. Detailed instructions for preparing and implementing a naive bayes classifier can be found in the Qiime2 documentation.
5. As the availability of *Fusobacterium* genome sequence data increases, we encourage the investigator to carefully choose representative *znpA* sequences accordingly. If multiple representative sequences from the same *F. nucleatum* subspecies are available, the user may consider including both sequences in the classifier or creating a subspecies-specific consensus sequence.
6. The user may want to modify input values for "—p-perc-identity." We recommend a value of 0.7, which we experimentally validated to yield nearly identical subspecies-level composition in samples compared to a separate methodology. However, the experimenter can modify this parameter to adjust stringency according to a desired experimental design.

Conflict of Interest Statement

The authors have no conflicts of interest to declare that are relevant to the content of this chapter.

References

1. Dzink JL, Sheenan MT, Socransky SS (1990) Proposal of three subspecies of *Fusobacterium nucleatum* Knorr 1922: *Fusobacterium nucleatum* subsp. *nucleatum* subsp. nov., comb. nov.; *Fusobacterium nucleatum* subsp. *polymorphum* subsp. nov., nom. rev., comb. nov.; and *Fusobacterium nucleatum* subsp. *vincentii* subsp. nov., nom. rev., comb. nov. Int J Syst Bacteriol

40:74–78. https://doi.org/10.1099/00207713-40-1-74

2. Gharbia SE, Shah HN, Lawson PA, Haapasalo M (1990) Distribution and frequency of *Fusobacterium nucleatum* subspecies in the human oral cavity. Oral Microbiol Immunol 5:324–327. https://doi.org/10.1111/j.1399-302x.1990.tb00434.x
3. Kook J-K, Park S-N, Lim YK et al (2017) Genome-based reclassification of *Fusobacterium nucleatum* subspecies at the species level. Curr Microbiol 74:1137–1147. https://doi.org/10.1007/s00284-017-1296-9
4. Bi D, Zhu Y, Gao Y et al (2022) Profiling *Fusobacterium* infection at high taxonomic resolution reveals lineage-specific correlations in colorectal cancer. Nat Commun 13:3336. https://doi.org/10.1038/s41467-022-30957-6
5. Krieger M, AbdelRahman YM, Choi D et al (2024) Stratification of *Fusobacterium nucleatum* by local health status in the oral cavity defines its subspecies disease association. Cell Host Microbe. https://doi.org/10.1016/j.chom.2024.02.010
6. Ye X, Wang R, Bhattacharya R et al (2017) *Fusobacterium nucleatum* subspecies *animalis* influences proinflammatory cytokine expression and monocyte activation in human colorectal tumors. Cancer Prev Res (Phila) 10:398–409. https://doi.org/10.1158/1940-6207.CAPR-16-0178
7. Senthakumaran T, Moen AEF, Tannæs TM et al (2023) Microbial dynamics with CRC progression: a study of the mucosal microbiota at multiple sites in cancers, adenomatous polyps, and healthy controls. Eur J Clin Microbiol Infect Dis. https://doi.org/10.1007/s10096-023-04551-7
8. Qin Y, Tong X, Mei W-J et al (2024) Consistent signatures in the human gut microbiome of old- and young-onset colorectal cancer. Nat Commun 15:3396. https://doi.org/10.1038/s41467-024-47523-x
9. Zepeda-Rivera M, Minot SS, Bouzek H et al (2024) A distinct *Fusobacterium nucleatum* clade dominates the colorectal cancer niche. Nature:1–9. https://doi.org/10.1038/s41586-024-07182-w
10. Kook J-K, Park S-N, Lim YK et al (2021) Correction to: genome-based reclassification of *Fusobacterium nucleatum* subspecies at the species level. Curr Microbiol 79:2. https://doi.org/10.1007/s00284-021-02680-w
11. Cho E, Park S-N, Shin Y et al (2015) *Peptoniphilus mikwangii* sp. nov., isolated from a clinical specimen of human origin. Curr Microbiol 70:260–266. https://doi.org/10.1007/s00284-014-0712-7
12. PJ MM, Holmes S (2013) phyloseq: an R package for reproducible interactive analysis and graphics of microbiome census data. PLoS One 8:e61217. https://doi.org/10.1371/journal.pone.0061217

Chapter 3

A Droplet Digital PCR Approach for the Quantitative Detection of *Fusobacterium nucleatum* in Formaldehyde-Fixed, Paraffin-Embedded Tumor Tissues

Orsalem Kahsai, Amanda I. Phipps, Polly A. Newcomb, and Meredith A. J. Hullar

Abstract

Fusobacterium nucleatum, a prevalent component of the intratumoral microbiome, has been associated with reduced survival in colorectal cancer. To enable sensitive and specific detection of *F. nucleatum* in tumor specimens, we developed a droplet digital PCR (ddPCR) assay targeting the transcription termination/anti-termination gene *nusG*, normalized to host tissue content using the solute carrier organic anion transporter family member 2A1 gene, SLCO2A1. Here, we present a ddPCR protocol optimized for quantifying *F. nucleatum* DNA in human genomic DNA extracted from formalin-fixed, paraffin-embedded (FFPE) tumor tissues. This methodology has been refined for high-throughput application, supporting large-scale analyses of *F. nucleatum* prevalence in tumor samples.

Key words Digital droplet PCR (ddPCR) quantitation, *Fusobacterium nucleatum animalis*, Intratumoral microbiome, Tumor microenvironment, CRC

1 Introduction

An integral part of the intratumoral microbiome, *Fusobacterium nucleatum*, has been associated with poorer survival outcomes from colorectal cancer (CRC) [1]. Detecting *F. nucleatum* in archived formalin-fixed, paraffin-embedded (FFPE) tumor tissues enables population-scale investigations that may improve risk stratification and inform therapeutic decision-making for CRC patients. However, accurate quantification of *F. nucleatum* in FFPE samples remains challenging due to the bacterium's low abundance in human tissues and the potential for extraneous environmental or paraffin-associated contamination that can obscure true microbial signal.

To address these challenges, we developed a droplet digital PCR (ddPCR) assay targeting the *nusG* gene, which encodes a

Hung Ton-That (ed.), *Fusobacterium nucleatum: Methods and Protocols*, Methods in Molecular Biology, vol. 3055,
https://doi.org/10.1007/978-1-0716-5388-3_3,

transcription termination/anti-termination factor in *F. nucleatum*. Quantification is normalized to host tissue content using the *SLCO2A1* gene, coding for the solute carrier organic anion transporter family member 2A1 [2]. Although this protocol was optimized for detecting *F. nucleatum* in CRC, the same principles can be adapted to detect other bacterial species in diverse tumor types.

Droplet digital PCR is a third-generation PCR technology that enables absolute quantitation of nucleic acid targets without the need for external calibration standards. Unlike quantitative PCR (qPCR), ddPCR is less susceptible to inhibition by environmental contaminants and matrix components common in FFPE-derived DNA [3]. Through microfluidic partitioning, ddPCR segregates each sample into thousands of nanoliter-sized droplets, within which PCR amplification occurs independently [4–7]. This partitioning greatly enhances sensitivity for detecting low-abundance targets in backgrounds dominated by human genomic DNA, such as FFPE tumor extracts [8–11].

Following amplification, droplets are individually analyzed and classified as positive or negative for the target sequence, and absolute copy number is calculated using Poisson statistics. Compared with qPCR, ddPCR offers several advantages, including improved analytical sensitivity [12], absolute quantification without standard curves, reduced susceptibility to PCR inhibitors [11, 13], lower variation in amplification efficiency, enhanced precision, and excellent reproducibility across runs [14, 15]. Moreover, ddPCR can detect subtle differences in target abundance with greater statistical power [16].

Here, we present a ddPCR protocol for the sensitive and specific quantification of Fn in genomic DNA extracted from FFPE CRC tumor tissues [2]. This optimized assay enables high-throughput, reproducible detection of intratumoral Fn and supports large-scale studies investigating microbial contributions to colorectal tumorigenesis and patient outcomes.

2 Materials

2.1 ddPCR Reaction Setup and Sample Preparation

1. Consumable plastics, reagents, and instrumentation required for the PCR reactions: a BioRad QX200™ system, ddPCR Supermix for Probes (no UTP), semi-skirted ddPCR 96-well plates, pierceable PCR plate heat-seal foil, filtered pipette tips for all pipetting steps, and an 8-channel multipipettor.
2. Deep well block PCR machine for thermal cycling of 96-well plates.
3. 96-well PCR plate sealer to seal plate with pierceable foil heat seal (5 s at 180 °C).

4. PCR primers and probes for the human reference gene and bacterial target:
5. Human reference: the *SLCO2A1* gene [2].

   F1 (forward) primer: 5′- ATCCCCAAAGCACCTGGTTT.

   R1 (reverse) primer: 5′-AGAGGCCAAGATAGTCCTGGTAA.

   P1 (minor groove binder (MGB) probe): 5′-CCATC-CATGTCCTCATCTC (6-VIC dye); VIC = 2′-chloro-7′-phenyl-1,4-dichloro-6-carboxyfluorescein.

   Bacterial target: *F. nucleatum nusG* [2].

   F1 (forward) primer: 5′-TGGTGTCATTCTTCCAA-AAATATCA.

   R1 (reverse) primer: 5′-AGATCAAGAAGGACAA-GTTGCTGAA.

   P1 (minor groove binder (MGB) probe): 5′-ACTTTAAC-TCTACCATGTTCA (5(6)-.

   Carboxyfluorescein FAM dye).

2.2 Load Sample Plate/Droplet Generation

1. For this protocol, a BioRad QX200™ droplet generator is used.
2. Automated droplet generator oil for probes.
3. DG32 automated droplet generator cartridges.
4. Pipet tips for the autogas system.
5. Pierceable PRC plate heat seal foils.
6. 96-well PCR plate sealer to seal plate with pierceable foil heat seal (red line up, 5 s at 180 °C).
7. Plate centrifuge.

2.3 PCR

1. Thermal cycler with 96-deep-well reaction module and heated lid.

2.4 Setup QX200™ Droplet Reader

1. QX200™ droplet reader (*see* **Note 1**).
2. ddPCR droplet reader oil.

2.5 Setup QuantaSoft™ Software to Run QX200™ Droplet Reader

1. Download software https://www.bio-rad.com/en-us/product/qx200-droplet-digital-pcr-system (v.1.7.4) on the laptop connected to QX200™ Droplet Reader.

2.6 Droplet Analysis Using Quantasoft™ Analysis Pro software

1. Download software: https://www.bio-rad.com/en-us/product/qx200-droplet-digital-pcr-system (click download under the main image) QuantaSoft™ Analysis Pro Software: pages 17–22.

3 Method

3.1 ddPCR Reaction Setup and Sample Preparation

1. Thaw all components to room temperature.
2. Mix thoroughly by vortexing (15 s at full speed).
3. Briefly centrifuge the tubes to collect all the reagents at the bottom of the tube.
4. Prepare a PCR master mix containing 11 ul of 2X ddPCR super mix for Probes (No dUTP):

   1.1 μL of forward and reverse primers (900 nM of final concentration each).

   1.1 μL of target and reference probes (the final concentration for FAM labeled probe is 250 nM and 62.5 nM for VIC-labeled probe).

   10 ng of genomic DNA.

   RNAase/DNAase-free water to a final reaction volume of 22 μL (*see* **Notes 2–4**).

3.2 Load Sample Plate/Generate Droplets (see Note 5)

1. Aliquot 10 μL of gDNA (1 ng/ μL) into a 96-well plate (layout samples in a predesigned 96-well plate map).
2. Using an 8-channel multichannel pipette, transfer 82 μL of master mix into 8 wells in a column of a separate 96-well plate.
3. Using multichannel pipet (eight channel), add 12 μL of Master Mix into all reaction wells of a 96-well ddPCR plate (*see* **Note 6**).
4. Mix the reaction very well by pipetting up and down.
5. Seal the plate using the Bio-Rad PX1 PCR Plate Sealer at 180 °C for 5 s, followed by vigorous and uniform mixing by vortexing the entire plate.
6. Briefly centrifuge the plate on a plate centrifuge for approximately 1 min at a low setting (i.e., 212 xG) or until most of the bubbles are eliminated and all reagents and DNA coalesce at the bottom of the tube.
7. Load consumables into the auto droplet generator (tips, oil, DG32, waste bin).
8. Generate droplets using the Bio-Rad Automatic Droplet Generator (AutoDG™) by loading the 96-well ddPCR plate with genomic DNA and master ix into the pre-chilled (−20 °C) droplet plate holder.
9. Immediately run the AutoDG™ protocol as per QX200 AutoDG™ manual.
10. When the droplet generation is finished, re-seal plate with another pierceable foil heat seal (red line up, 5 s 180 °C). Sealed plate may be stored in the dark for up to 1 h at room temperature or up to 4 h at 4 °C.

3.3 PCR

1. After droplet generation is complete (*see* Subheading 3.2, **step 10**), gently transfer the sealed 96-well plate to the deep well thermocycler.
2. Perform PCR using the following conditions to generate reference and target amplicons (*see* **Note 7**):

   Step 1 (one cycle, DNA denaturation/DNA polymerase activation).

   95 °C, 10 min.

   Step 2 (50 cycles).

   95 °C, 30 s at ramp rate of 2 °C/s (denaturation).

   60 °C, 2 min at a ramp rate of 2 °C/s (annealing/extension ~ T_m of primers and probes).

   Step 3 (one cycle: enzyme deactivation post-cycling steps).

   98 °C for 10 min at a ramp rate of 2 °C/s.

   Step 5 (Hold).

   10 °C.

3.4 Setup QX200 Droplet Reader (see Note 8)

1. Remove the 96-well plate holder from the QX200 Droplet Reader, and take the top silver part off by unclipping the sides.
2. Take the 96-well sample plate out of the thermocycler slowly to not disturb the droplets, and place into the black plate holder.
3. Gently put the silver plate top on the 96-well sample plate, and while holding down the top part, place the side locks down gently.
4. Place the 96-well plate holder containing the 96-well PCR plate into the reader gently.
5. Double check the two bins on the side of the reader to ensure there is no need for changing.

3.5 Running the QX200 Droplet Reader Using QuantaSoft™ Software (see Note 9)

1. Open **QuantaSoft™** version 1.7.4 on the laptop attached to the reader.
2. In the **QuantaSoft™** software, click on the FLUSH button found on the side of the software screen to flush the QX200 reader system.
3. In the **QuantaSoft™** software, click on the PRIME button found on the side of the software screen to flush the QX200 Droplet Reader system. The QX200 Droplet Reader is now ready to use.
4. Double click on any well in the Plate Map to enter the Plate Editor.

5. Select all reaction wells/columns (Shift+click for contiguous wells or Ctrl+click for noncontiguous wells).
6. Under *Sample*, select the Experiment type from the menu (if in doubt, select ABS).
7. Under *Sample*, select the EXACT Super Mix name (ddPCR Super Mix for probes (no dUTP)).
8. Under *Target 1*, add name of FAM target (in this case; FAM Fn *nusG*).
9. Under *Target 1*, change Type from Unused to Unknown.
10. Under *Target 2*, add name of HEX/VIC target (in this case; VIC SCLOA1).
11. Under *Target 2*, change Type from Unused to Unknown (if applicable).
12. Click on Apply (or tap enter).
13. Now, select each well (or wells if replicates), and under *Sample*, enter sample name.
14. Click on Apply (or tap enter).
15. Repeat Steps 13 and 14 for each sample.
16. Click *OK* to confirm changes.
17. Click on the *Run* tab (confirm the prompt and enter a run name).
18. Verify Dye Set (FAM and/or VIC).
19. Click OK to start (*see* **Note 10**).

3.6 QX200™ Data Analysis

1. Upon completion of the QX200™ Droplet Reader run, save the.qlp data file to a thumb drive for transfer.
   (a) Under *Type*, file is labeled QuantaSoft Plate.
2. Open file in QuantaSoft™ *Analysis Pro.*
3. Verify the number of droplets.
   (a) Select all reaction wells in the Plate Map.
   (b) Click on the *Droplets* tab.
   (c) Confirm all wells >10 K droplets.
4. Verify successful amplification.
   (a) Click on *1D Amplitude* (for singleplex) or *2D Amplitude* (for multiplex).
   (b) Verify presence and separation of positive and negative droplets in positive control sample.
   (c) Verify the absence of positive droplets in negative control sample/NTC.

5. Set threshold.
   (a) Select all wells containing the same assay and fluorescence channel.
   (b) Under Graph Tools, select line (1D amplitude view) or crosshair (2D cluster view).
   (c) Click on the threshold line in the space between the negative and positive droplet clusters (bands).
   (d) Adjust the threshold, by moving the cursor over the pink threshold line, then click and drag the line up or down until the Positive and negative droplets are clearly separated.
6. View data.
   (a) Click on the *Concentration* tab.
   (b) Adjust graph options to your preference.
   (c) Export data table by right clicking and selecting Export preference.
   (d) The counts of *Fn* detected in tissue samples are normalized for the amount of human tissue added to each sample using the counts of the eucaryotic housekeeping gene, SLCO2A1 (e.g., the counts of Fn(*nus*G) divided by the counts of SLCO2A1).

4 Notes

1. Droplet Reader maintenance: Prime the droplet reader system if it has been idle for more than 5 days (see Manual). When replacing the waste bottle, add 50 mL of 10% bleach. Ensure the inside, outside, and lid of the waste container are clean. If reusing the waste bottle, rinse thoroughly with warm water. Power cycle the reader and laptop monthly.
2. The ddPCR reactions were optimized for 10 ng of genomic DNA extracted from FFPE tumor tissue, which contained mixed human and bacterial DNA. Add 10 μL of sample at 1 ng/μL genomic DNA.
3. Prepare the master mix based on the total number of samples, including positive, negative, and wax controls, plus one extra reaction to account for pipetting error.
4. For 96-well high-throughput ddPCR, first prepare the master mix without genomic DNA. Add genomic DNA, controls, or standards directly to each well to reduce pipetting variability, minimize contamination risk, and ensure consistent reaction conditions. Dispense the remaining master mix uniformly across the plate using multiple pipette tips.

5. Load sample plate within 45 min of preparing the master mix.
6. Each reaction consists of 10 μL of genomic DNA and 12 μL of master mix (total 22 μL). The Supermix must account for 50% of the final reaction volume.
7. PCR thermal cycling: Optimize the PCR annealing temperature within 55–65 °C—this is essential. For assays adapted from real-time qPCR, decrease the annealing temperature by 2 °C. For amplicons longer than 250 bp, extend the annealing step to 2 min. For extended holds, set the temperature to 12 °C to avoid condensation.
8. Our standard operating procedure is to read the plate within 1 h after amplification. However, plates may be stored at 4 °C in the dark for up to 3 days; allow them to reach room temperature before reading.
9. The information is summarized in the manual for the QX200™ Droplet Reader and QuantaSoft™ software (pages 9–13).
10. Refer to Table 1, Chapter 2 in the QX200™ manual for explanations of droplet reader status indicator lights if a run does not progress.

Acknowledgments

We would like to thank our laboratory members for their critical review of this manuscript. Additionally, we would like to thank Dr. Shuji Ogino for helpful scientific discussions and Rachel Malen for her excellent project management. This work was supported by the NCI, NIH grants U01CA74794 (to P.A. Newcomb), U01CA167551 (to A.I. Phipps), R01CA076366 (to P.A. Newcomb), R01CA217970 (to A.I. Phipps, P.A. Newcomb, and M.A.J. Hullar), and P50CA285275 (to A.I. Phipps and M.A.J. Hullar). The content presented in this publication does not reflect the viewpoints of the National Institutes of Health and is solely the responsibility of the authors.

Conflict of Interest Statement The authors have no conflicts of interest to declare that are relevant to the content of this chapter.

References

1. Phipps AI, Hill CM, Lin G, Malen RC, Reedy AM, Kahsai O, Ammar H, Curtis K, Ma N, Randolph TW, Ma J, Ogino S, Newcomb PA, Hullar MA (2025) Fusobacterium nucleatum enrichment in colorectal tumor tissue: associations with tumor characteristics and survival outcomes. Gastro Hep Adv 4(6):100644. https://doi.org/10.1016/j.gastha.2025.100644

2. Hullar MA, Kahsai OJ, Hill C, Levy L, Malen RC, Curtis KR, Ammar H, Sillah A, Reedy AM, Lampe JW, Ogino S, Potter JD, Newcomb PA, Phipps AI (2025) Highly sensitive DNA testing of Fusobacterium nucleatum (Fn) in colorectal tumors. Cancer Epidemiol Biomarkers Prev:1377–1385. https://doi.org/10.1158/1055-9965.EPI-24-1020
3. Miotke L, Lau BT, Rumma RT, Ji HP (2014) High sensitivity detection and quantitation of DNA copy number and single nucleotide variants with single color droplet digital PCR. Anal Chem 86(5):2618–2624. https://doi.org/10.1021/ac403843j
4. Robins HS, Ericson NG, Guenthoer J, O'Briant KC, Tewari M, Drescher CW, Bielas JH (2013) Digital genomic quantification of tumor-infiltrating lymphocytes. Sci Transl Med 5(214):214ra169. https://doi.org/10.1126/scitranslmed.3007247
5. Nakano M, Komatsu J, Matsuura S, Takashima K, Katsura S, Mizuno A (2003) Single-molecule PCR using water-in-oil emulsion. J Biotechnol 102(2):117–124. https://doi.org/10.1016/s0168-1656(03)00023-3
6. Hindson BJ, Ness KD, Masquelier DA, Belgrader P, Heredia NJ, Makarewicz AJ, Bright IJ, Lucero MY, Hiddessen AL, Legler TC, Kitano TK, Hodel MR, Petersen JF, Wyatt PW, Steenblock ER, Shah PH, Bousse LJ, Troup CB, Mellen JC, Wittmann DK, Erndt NG, Cauley TH, Koehler RT, So AP, Dube S, Rose KA, Montesclaros L, Wang SL, Stumbo DP, Hodges SP, Romine S, Milanovich FP, White HE, Regan JF, Karlin-Neumann GA, Hindson CM, Saxonov S, Colston BW (2011) High-throughput Droplet Digital PCR system for absolute quantitation of DNA copy number. Anal Chem 83(22):8604–8610. https://doi.org/10.1021/ac202028g
7. Pinheiro LB, Coleman VA, Hindson CM, Herrmann J, Hindson BJ, Bhat S, Emslie KR (2012) Evaluation of a droplet digital polymerase chain reaction format for DNA copy number quantification. Anal Chem 84(2): 1003–1011. https://doi.org/10.1021/ac202578x
8. Talarico S, Safaeian M, Gonzalez P, Hildesheim A, Herrero R, Porras C, Cortes B, Larson A, Fang FC, Salama NR (2016) Quantitative detection and genotyping of Helicobacter pylori from stool using droplet digital PCR reveals variation in bacterial loads that correlates with cagA virulence gene carriage. Helicobacter 21(4):325–333. https://doi.org/10.1111/hel.12289
9. Sze MA, Abbasi M, Hogg JC, Sin DD (2014) A comparison between droplet digital and quantitative PCR in the analysis of bacterial 16S load in lung tissue samples from control and COPD GOLD 2. PLoS One 9(10): e110351. https://doi.org/10.1371/journal.pone.0110351
10. Qian L, Song H, Cai W (2016) Determination of Bifidobacterium and lactobacillus in breast milk of healthy women by digital PCR. Benef Microbes 7(4):559–569. https://doi.org/10.3920/BM2015.0195
11. Sedlak RH, Kuypers J, Jerome KR (2014) A multiplexed droplet digital PCR assay performs better than qPCR on inhibition prone samples. Diagn Microbiol Infect Dis 80(4):285–286. https://doi.org/10.1016/j.diagmicrobio.2014.09.004
12. Ma J, Li N, Guarnera M, Jiang F (2013) Quantification of plasma miRNAs by digital PCR for cancer diagnosis. Biomark Insights 8:127–136. https://doi.org/10.4137/BMI.S13154
13. Dingle TC, Sedlak RH, Cook L, Jerome KR (2013) Tolerance of droplet-digital PCR vs real-time quantitative PCR to inhibitory substances. Clin Chem 59(11):1670–1672. https://doi.org/10.1373/clinchem.2013.211045
14. Mukaide M, Sugiyama M, Korenaga M, Murata K, Kanto T, Masaki N, Mizokami M (2014) High-throughput and sensitive next-generation droplet digital PCR assay for the quantitation of the hepatitis C virus mutation at core amino acid 70. J Virol Methods 207: 169–177. https://doi.org/10.1016/j.jviromet.2014.07.006
15. Hindson CM, Chevillet JR, Briggs HA, Gallichotte EN, Ruf IK, Hindson BJ, Vessella RL, Tewari M (2013) Absolute quantification by droplet digital PCR versus analog real-time PCR. Nat Methods 10(10):1003–1005. https://doi.org/10.1038/nmeth.2633
16. Bharuthram A, Paximadis M, Picton AC, Tiemessen CT (2014) Comparison of a quantitative Real-Time PCR assay and droplet digital PCR for copy number analysis of the CCL4L genes. Infect Genet Evol 25:28–35. https://doi.org/10.1016/j.meegid.2014.03.028

Chapter 4

Enhanced Transformation Efficiency in *Fusobacterium nucleatum* by Bypassing Restriction-Modification Systems: A Methylation-Guided Protocol

Ling Liu, Yu Zhang, and Xue Liu

Abstract

The genetic intractability of *Fusobacterium nucleatum* has slowed progress in defining mechanisms of fusobacterial pathogenesis. Transformation is inefficient, and broadly applicable genetic tools are still limited. A major obstacle is the presence of endogenous restriction-modification (RM) systems that cleave incoming DNA lacking appropriate methylation. This chapter presents a concise, reproducible protocol to increase transformation efficiency in *F. nucleatum* by systematically bypassing RM barriers. The workflow combines methylome mapping by single-molecule real-time (SMRT) sequencing, expression of cognate methyltransferases (MTases) in an intermediary *Escherichia coli* host to pre-methylate shuttle plasmids, and optimized electroporation with careful anaerobic handling and selective recovery. Using *F. nucleatum* ATCC 25586 as an example, the method yields a marked improvement in transformation efficiency, and the same logic can be adapted to additional *F. nucleatum* isolates once their methylation context is defined.

Key words *Fusobacterium nucleatum*, Restriction-modification, Transformation, Methylation

1 Introduction

F. nucleatum is a Gram-negative, spindle-shaped, obligate anaerobe that colonizes human mucosal surfaces, particularly the oral cavity and gastrointestinal tract. It acts as a bridging organism in dental biofilms and has been implicated in periodontitis, invasive head-and-neck infections, adverse pregnancy outcomes, and colorectal cancer [1]. The causal links between fusobacterial factors and disease remain incompletely defined, in part because genetic manipulation has been successful in only a few strains of *F. nucleatum* [2–4]. Bacteria deploy multiple innate and adaptive defense systems against foreign DNA [5–8]. Among these, restriction-modification (RM) systems are pervasive,being present in roughly 95% of

Ling Liu and Yu Zhang contributed equally to this work.

Hung Ton-That (ed.), *Fusobacterium nucleatum: Methods and Protocols*, Methods in Molecular Biology, vol. 3055,
https://doi.org/10.1007/978-1-0716-5388-3_4,

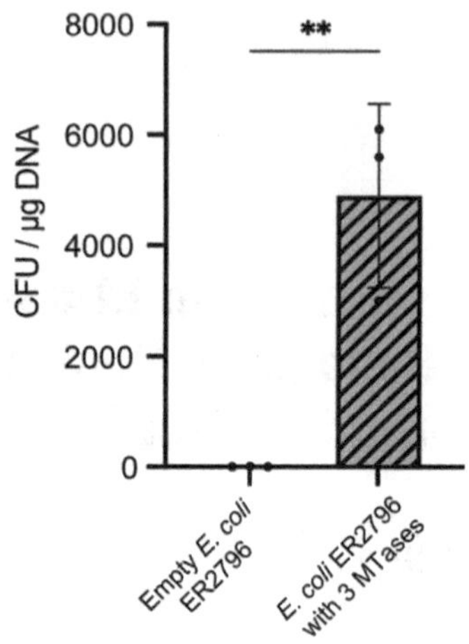

Fig. 1 Transformation efficiency of *F. nucleatum* ATCC 25586 using the shuttle plasmid pFN1-catP. Shuttle plasmids were prepared in *E. coli* ER2796 engineered to express the indicated DNA methyltransferases (MTases) to confer host-like methylation patterns prior to transformation. Methylated plasmids lead to significantly increased transformation efficiency ($p < 0.01$)

sequenced bacterial genomes and constitute a major barrier to DNA transfer [9]. This context suggests that the low transformation efficiency of *F. nucleatum* is driven by active RM systems. A recent analysis predicted diverse RM system repertoires across *Fusobacterium* species and strains [10].

Here, we use SMRT (PacBio) sequencing to define methylated DNA motifs in *F. nucleatum* and develop a practical pipeline that circumvents RM-mediated restriction. We identify candidate MTases from the methylome and REBASE prediction, express the active MTases in *E. coli* to pre-methylate plasmids, and transform the methylated plasmids into *F. nucleatum* under rigorously controlled anaerobic conditions. This procedure enables high-efficiency transformation under our experimental conditions (*see* Fig. 1) and is designed to be transferable to additional *F. nucleatum* isolates (*see* Fig. 2).

2 Materials

Prepare all solutions with sterile, molecular-grade water. Handle *F. nucleatum* in an anaerobic chamber whenever possible.

2.1 Identification of DNA Methylation Motifs and Cognate Methyltransferases

1. Strain: *F. nucleatum* ATCC 25586.
2. Brain heart infusion (BHI) medium.
3. 1.5-mL Eppendorf tubes.

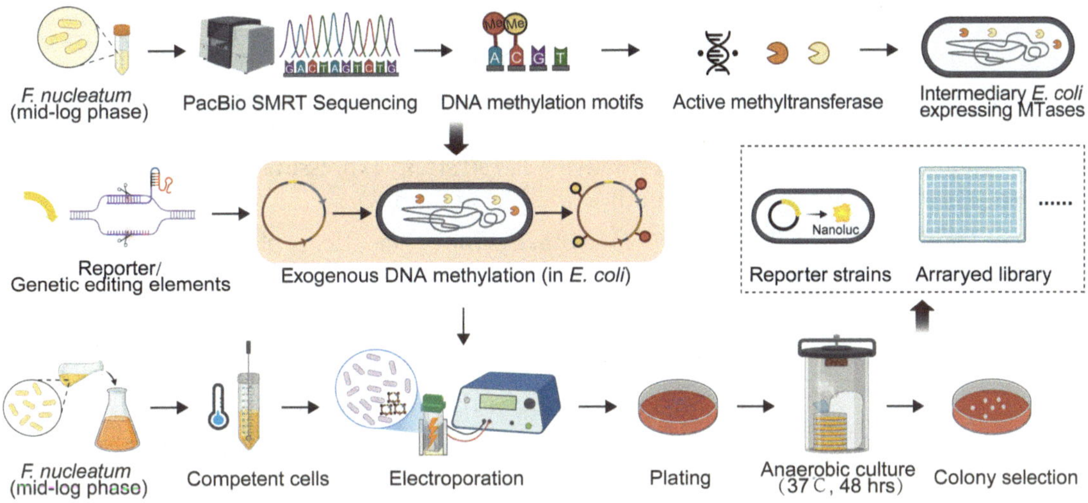

Fig. 2 Integrated workflow to enhance transformation efficiency in *F. nucleatum* by circumventing restriction-modification (R-M) barriers (Created in BioRender. Liu, X. (2025) https:/BioRender.com/zbdg67d). The approach profiles the operative R-M systems in the target strain, engineers recombinant vectors bearing host-like methylation patterns to evade endonuclease cleavage, and optimizes DNA-delivery parameters (e.g., electroporation) to maximize uptake of these "stealth" constructs. This strain-aware, integrated strategy substantially increases transformation success in *F. nucleatum* strains, thereby facilitating genetic manipulation and functional studies of fusobacterial biology and pathogenesis

2.2 Construction of an Intermediary E. coli Strain Harboring F. nucleatum Methyltransferases

1. Luria-Bertani (LB) broth and LB agar.
2. Antibiotics for *E. coli*: 20 μg/mL gentamicin; 100 μg/mL ampicillin as final concentration.
3. 30 mM 2,6-diaminopimelic acid (DAP) stock solution in water; use at a final concentration of 300 μM.
4. Host *E. coli* for MTase expression: *E. coli* ER2796 [11], with DNA methyltransferase-deficient background, was used for heterologous expression of *F. nucleatum* MTases (*see* **Note 1**).
5. *E. coli* strain for hosting Tn7 integration and helper plasmids: *E. coli* WM3064, a widely used conjugation donor strain, was employed for mobilization of Tn7-based constructs (*see* **Note 2**).
6. Tn7 integration plasmid (obtained from Addgene #119278) and Tn7 helper plasmids (obtained from Addgene #119239).
7. High-fidelity DNA polymerase.
8. Genomic DNA isolation kit.
9. Plasmid miniprep kit.
10. Agarose.
11. TAE buffer (40 mM of Tris base, 20 mM acetic acids, 1 mM EDTA; pH 8.0).
12. Nuclease-free water.
13. Sterile PBS (10 mM of Na_2HPO_4, 1.8 mM of KH_2PO_4, 137 mM of NaCl, 2.7 mM of KCl; pH 7.4).

14. Sterile razor blade for gel excision.
15. Orbital shaker incubator.
16. Benchtop centrifuge.

2.3 Preparation of Pre-methylated Plasmids Tailored for F. nucleatum *Transformation*

1. The intermediary *E. coli* strain expressing *F. nucleatum* MTases (from methods Subheading 3.2).
2. *E. coli-F. nucleatum* shuttle plasmid (pFN1-catP, Addgene #244451).
3. Ice and pre-chilled sterile water.
4. LB agar with 20 μg/mL gentamicin.
5. LB agar with 15 μg/mL chloramphenicol.
6. LB agar with 300 μM DAP and 20 μg/mL gentamicin (as required).
7. Spectrophotometer (e.g., NanoDrop).
8. 50-mL centrifuge tubes with caps.
9. 2-mL cryovials.

2.4 Generation of Competent F. nucleatum *Cells and Optimized Electroporation-Based DNA Delivery*

1. *F. nucleatum* ATCC 25586.
2. Pre-methylated plasmid pFN1-catP.
3. LB agar with 15 μg/mL chloramphenicol for selection of *E. coli* transformants.
4. Columbia blood agar (CBA) with 5% sheep blood and 5 μg/mL thiamphenicol for selection of *F. nucleatum* transformants.
5. 1.5-mL Eppendorf tubes.
6. 0.2-mL PCR tubes.
7. 15-mL centrifuge tubes with caps.
8. Standard laboratory pipettes and sterile tips.
9. Sterile spreaders.
10. Electroporator with 2-mm gap cuvettes.
11. Anaerobic chamber (5% CO_2, 5% H_2, 90% N_2).
12. Ice and pre-chilled sterile water.
13. High-fidelity DNA polymerase.

3 Methods

3.1 Identification of DNA Methylation Motifs and Cognate Methyltransferases

1. Identify methylated DNA motifs in *F. nucleatum* ATCC 25586 using a commercial PacBio SMRT sequencing service (*see* **Note 3**).
2. Integrate modification calls with enzyme/motif annotations in REBASE (http://rebase.neb.com/cgi-bin/pacbiolist) to nominate candidate MTases (*see* **Note 4**).

3.2 Construction of an Intermediary *E. coli* Strain Harboring *F. nucleatum* Methyltransferases

This part consists of three major steps: design and synthesis of MTase gene blocks, assembly of these genes into Tn7 integration vector, and integration of the MTase expression cassette into the chromosome of *E. coli*.

1. Codon-optimize the candidate MTase genes from *F. nucleatum* ATCC 25586 for expression in *E. coli* ER2796 and synthesize them as double-stranded DNA fragments (e.g., gBlocks) (*see* **Notes 5–6**).
2. Amplify synthesized fragments by PCR and assemble into the Tn7 integration vector using Gibson Assembly or an equivalent method.
3. Introduce the resulting Tn7 integration vector carrying the MTase expression cassette into *E. coli* WM3064, which serves as the donor strain in the subsequent conjugation step (*see* **step 5**).
4. In parallel, introduce the Tn7 helper plasmid into *E. coli* WM3064, and the resulting strain will serve as the helper strain for the next step (*see* **step 5**).
5. Inoculate single colonies of the donor strain (**step 3**), the helper strain (**step 4**), and the recipient strain *E. coli* ER2796 into LB medium supplemented with appropriate antibiotics. Add 300 μM DAP only to cultures of *E. coli* WM3064-derived strains.
6. Incubate the cultures overnight at 37 °C with shaking at 200 rpm.
7. The next day, transfer approximately 50 μL of each overnight culture into 15 mL fresh LB. Supplement donor and helper strain cultures with the appropriate antibiotics and 300 μM DAP. Supplement the recipient *E. coli* ER2796 culture with 100 μg/mL ampicillin but without DAP.
8. Incubate the fresh cultures at 37 °C with shaking at 200 rpm for 4 h to mid-exponential phase (OD_{600} between 0.4 and 0.6).
9. Harvest the cells by centrifugation at 1800 × g for 5 min at 4 °C, wash once with 5 mL of LB, and resuspend in 15 mL of fresh LB.
10. Combine the resuspended helper, donor, and recipient strains at a 1:1:1 ratio (100 μL each). Pellet by centrifugation at 7000 × g for 2 min, resuspend the mixed pellet in 30 μL LB, and spot it onto LB agar containing 300 μM DAP for conjugation.
11. After incubation at 37 °C overnight (~18 h), resuspend the mating lawn in 900 μL of PBS. Prepare serial dilutions and plate 100 μL of the dilutions onto LB agar supplemented with 20 μg/mL gentamicin to select for successful transconjugants.

12. Confirm transconjugants by colony PCR and Sanger sequencing. Stock successfully validated transconjugants and use them as the intermediary *E. coli* host for plasmid pre-methylation in the following subheading.

3.3 Preparation of Pre-methylated Plasmids Tailored for F. nucleatum Transformation

1. Inoculate a single colony of the intermediary *E. coli* host strain (*see* Subheading 3.2, **step 12**) into 10 mL of LB supplemented with 20 μg/mL gentamicin and incubate overnight at 37 °C with shaking (200 rpm).
2. Transfer 1 mL of the overnight culture into 100 mL of fresh LB medium, and incubate at 37 °C with shaking for 4–5 h (OD_{600} between 0.4 and 0.6).
3. Split the 100-mL culture into two 50-mL centrifuge tubes, keep them on ice for 20 min, and then harvest the cells by centrifugation at 7000 × g for 5 min at 4 °C.
4. Wash the cell pellets three times with 25 mL of ice-cold sterile water, centrifuging between washes at 7000 × g for 5 min at 4 °C.
5. Resuspend the final pellets in 2 mL (per 50 mL tube) of ice-cold sterile water containing 10% glycerol to obtain electrocompetent cells.
6. Aliquot 200 μL of cell suspensions into 1.5-mL tubes, immediately snap-freeze in liquid nitrogen, and store at −80 °C for future usage.
7. Add ~300 ng plasmid DNA (pFN1-catP) into 100 μL of competent cells (**step 6**), mix gently, and transfer to a pre-chilled 2-mm cuvette, avoiding bubbles.
8. Electroporate at 1.8 kV, immediately add 900 μL pre-warmed (37 °C) LB to the cuvette to initiate recovery, transfer to a new 1.5-mL tube, and incubate at 37 °C with shaking (200 rpm) for 2–3 h.
9. Plate 50 μL on LB agar plates containing 15 μg/mL chloramphenicol and incubate at 37 °C until colonies appear (up to 2 days for pFN1-catP).
10. Inoculate single colonies into selective LB, and grow overnight for preparation of plasmid DNA using a miniprep kit. The resulting plasmid DNA carries the methylation pattern conferred by the expressed F. nucleatum MTases and is used as pre-methylated plasmid DNA for subsequent transformation.

3.4 Generation of Competent F. nucleatum Cells and Optimized Electroporation-Based DNA Delivery

Procedures are adapted from a previously reported protocol [3] with minor modifications (*see* **Notes 7–8**).

1. Streak *F. nucleatum* ATCC 25586 from −80 °C stock onto CBA plates, and incubate anaerobically at 37 °C for 2–3 days to obtain single colonies.

2. Pool several colonies into 12-mL BHI, and incubate overnight at 37 °C in an anaerobic chamber.
3. Transfer the entire 12-mL culture into 40-mL fresh BHI in a 50-mL tube.
4. Incubate anaerobically at 37 °C for 4 days until OD_{600} ~ 1.2.
5. Chill on ice for 10–20 min, and harvest cells by centrifugation at 7000 × g for 5 min at 4 °C.
6. Wash the pellets three times with at least 25 mL of pre-chilled sterile water.
7. Resuspend in 1 mL of ice-cold 10% glycerol, aliquot 200 μL into 1.5-mL tubes, snap-freeze in liquid nitrogen, and store the aliquots at −80 °C for future usage.
8. Thaw one tube of competent cells on ice, add 15 μL (~1 μg) of pre-methylated plasmid (*see* Subheading 3.3, **step 10**) to 150 μL of competent cells, and incubate on ice for ≥10 min.
9. Transfer 150 μL of the mixture into a pre-chilled 2-mm cuvette and electroporate at 2.5 kV, 25 μF, 200 Ω.
10. Immediately add 1 mL of BHI to the cuvette, and transfer the suspension to a sterile 1.5-mL tube.
11. Incubate at 37 °C for 6 h in the anaerobic chamber.
12. Harvest the cell pellets by centrifugation at 7000 × g for 5 min at room temperature, resuspend the pellet in 100 μL of BHI medium, and plate the entire resuspension onto CBA plates with 5% sheep blood and 5 μg/mL thiamphenicol.
13. Incubate anaerobically at 37 °C for 3 days.
14. Confirm colonies by PCR and Sanger sequencing.

4 Notes

1. The *E. coli* ER2796 strain is used for heterologous expression of *F. nucleatum* MTases. Because this strain is DNA methyltransferase-deficient, it does not introduce any additional methylation beyond that provided by the *F. nucleatum* MTases themselves. Other *E. coli* strains, such as DH5α, can also be employed for MTase expression in cases where the methylation motifs of *F. nucleatum* MTases are not subject to interference by native *E. coli* MTases. For certain *F. nucleatum* strains, such as strain 10953, which utilize GATC N6-methyladenine (6mA) methylation, DH5α may even be preferable, since its native Dam activity supplies GATC (6mA) methylation directly.
2. The *E. coli* strain WM3064 is auxotrophic for diaminopimelic acid (DAP), a lysine biosynthesis precursor. This auxotrophy

allows efficient counter-selection and removal of the strain from bi- or tri-parental mating mixtures following conjugation. WM3064 is commonly used both as a conjugation donor and for maintenance of plasmids requiring the Pir protein for replication. The strain was originally derived from *E. coli* B2155 by William Metcalf (University of Illinois at Urbana-Champaign). Of note, the genotype of WM3064 is *thrB1004 pro thi rpsL hsdS lacZΔM15 RP4–1360 Δ(araBAD)567 ΔdapA1341::[erm pir]*.

3. DNA methylation motifs can be identified using either PacBio SMRT sequencing or, where appropriate basecalling and modification-detection workflows are available, Oxford Nanopore sequencing. The choice of platform can be made according to availability and project needs. For certain *F. nucleatum* strains, predicted MTases are already listed in the REBASE database, and these candidates can be directly selected for cloning, making SMRT sequencing optional. Nevertheless, sequencing remains the most reliable approach for comprehensive identification of all active DNA methylation motifs. In cases where specific DNA methylation motifs in *F. nucleatum* cannot be readily attributed to endogenous MTases, these motifs can be introduced using well-characterized MTases from other species.

4. The Rebase database (https://rebase.neb.com/rebase/rebase.html) is a comprehensive resource for restriction-modification systems, DNA methyltransferases, recognition motifs, and related enzymes.

5. Codon optimization is often required to ensure efficient heterologous expression of MTases or other proteins in *E. coli* or alternative hosts. Many commercial gene synthesis providers offer built-in codon optimization when designing synthetic DNA fragments (gBlocks). Examples include IDT codon optimization tool (Integrated DNA Technologies), Twist Bioscience codon optimization for gene fragments, and GenScript OptimumGene™ optimization algorithm.

6. For synthetic MTase constructs, it is essential to include a constitutive promoter to drive expression in *E. coli*. A commonly used example is the synthetic constitutive P3 promoter (5′-ccattctacagtttattcttgacattgcactgtccccctggtataataactata-3′). The promoter can be placed directly upstream of the coding sequence during gene synthesis. Alternative constitutive promoters may also be selected depending on the desired expression level and compatibility with the host strain. For expression of multiple MTase-encoding genes, the genes can be organized into a single operon and driven by the same promoter.

7. Because of oxygen sensitivity, *F. nucleatum* viability declines rapidly under aerobic conditions. Perform time-critical steps promptly and, where possible, inside the anaerobic chamber.
8. Plasmid stability: Isolate plasmids carrying synthetic MTase genes from intermediary hosts within two subculture cycles to preserve sequence integrity.

Acknowledgments

This work was supported by the National Key Research and Development Program of China (2023YFD10800100), the National Natural Science Foundation of China (82200047, 82270012), the Science and Technology Project of Shenzhen (JCYJ20220818095602006), Shenzhen Basic Research Program (General Program) (JCYJ20240813143700001), and the Shenzhen University 2035 Program for Excellent Research (86901-00000216).

Conflict of Interest Statement The authors have no conflicts of interest to declare that are relevant to the content of this chapter.

References

1. Ponath F, Zhu Y, Cosi V, Vogel J (2022) Expanding the genetic toolkit helps dissect a global stress response in the early-branching species *Fusobacterium nucleatum*. Proc Natl Acad Sci USA 119. https://doi.org/10.1073/pnas.2201460119
2. Nakagaki H, Sekine S, Terao Y, Toe M, Tanaka M, Ito H-O, Kawabata S, Shizukuishi S, Fujihashi K, Kataoka K (2010) *Fusobacterium nucleatum* envelope protein FomA is immunogenic and binds to the salivary statherin-derived peptide. Infect Immun 78:1185–1192. https://doi.org/10.1128/IAI.01224-09
3. Peluso EA, Scheible M, Ton-That H, Wu C (2020) Genetic manipulation and virulence assessment of *Fusobacterium nucleatum*. Curr Protoc Microbiol 57. https://doi.org/10.1002/cpmc.104
4. Yoshida A, Ikegami A (2021) Genetic transformation of *Fusobacterium nucleatum*. In: Methods in molecular biology, pp 43–50, https://doi.org/10.1007/978-1-0716-0939-2_5
5. Mohanraju P, Makarova KS, Zetsche B, Zhang F, Koonin EV, van der Oost J (2016) Diverse evolutionary roots and mechanistic variations of the CRISPR-Cas systems. Science 353:aad5147. https://doi.org/10.1126/science.aad5147
6. Hegge JW, Swarts DC, van der Oost J (2018) Prokaryotic Argonaute proteins: novel genome-editing tools? Nat Rev Microbiol 16: 5–11. https://doi.org/10.1038/nrmicro.2017.73
7. van Houte S, Buckling A, Westra ER (2016) Evolutionary ecology of prokaryotic immune mechanisms. Microbiol Mol Biol Rev 80:745–763. https://doi.org/10.1128/MMBR.00011-16
8. Goldfarb T, Sberro H, Weinstock E, Cohen O, Doron S, Charpak-Amikam Y, Afik S, Ofir G, Sorek R (2015) BREX is a novel phage resistance system widespread in microbial genomes. EMBO J 34:169–183. https://doi.org/10.15252/embj.201489455
9. Roberts RJ, Vincze T, Posfai J, Macelis D (2015) REBASE--a database for DNA restriction and modification: enzymes, genes and genomes. Nucleic Acids Res 43:D298–D299. https://doi.org/10.1093/nar/gku1046
10. Umaña A, Nguyen TTD, Sanders BE, Williams KJ, Wozniak B, Slade DJ (2022) Enhanced

Fusobacterium nucleatum genetics using host DNA methyltransferases to bypass restriction-modification systems. J Bacteriol 204. https://doi.org/10.1128/jb.00279-22

11. Anton BP, Mongodin EF, Agrawal S, Fomenkov A, Byrd DR, Roberts RJ, Raleigh EA (2015) Complete genome sequence of ER2796, a DNA methyltransferase-deficient strain of *Escherichia coli* K-12. PLoS One 10: e0127446. https://doi.org/10.1371/journal.pone.0127446

Chapter 5

A Conditional Plasmid System for Markerless Gene Deletion in Genetically Recalcitrant *Fusobacterium nucleatum* subsp. *animalis*

Bibek G C and Chenggang Wu

Abstract

Fusobacterium nucleatum subsp. *animalis* (FNA) plays a prominent role in oral microbial ecology and is increasingly implicated in systemic diseases such as colorectal cancer. However, its genetic intractability has significantly hindered functional studies to understand its pathogenic mechanisms. To address this challenge, we developed a conditional plasmid system that enables efficient, markerless gene deletion in FNA strains. This system features inducible control of plasmid replication via a theophylline-responsive riboswitch regulating *repA*, and counterselection via the MazF toxin under the control of an anhydrotetracycline-inducible promoter. In this chapter, we provide a detailed, step-by-step protocol for implementing this system using FNA strain 7_1 as a model. We illustrate the procedure by deleting the *luxS* gene, which encodes S-ribosylhomocysteine lyase—an enzyme involved in AI-2 quorum sensing and biofilm regulation. Our protocol provides a powerful and adaptable tool for advancing genetic studies in this genetically recalcitrant subspecies.

Key words *Fusobacterium nucleatum* subsp. *animalis*, Gene deletion, *LuxS*, Conditional plasmid, *mazF* as a counterselection marker

1 Introduction

Fusobacterium nucleatum is a Gram-negative, strictly anaerobic, spindle-shaped bacterium that produces hydrogen sulfide (H_2S), indole, and butyrate [1]. It is a key constituent of the oral microbiota and is particularly abundant in subgingival plaque, where it plays a central role in periodontal disease development [2]. Historically, research on *F. nucleatum* was centered in oral microbiology, focusing on its ecological role in dental health and disease. However, over the past decade, the significance of *F. nucleatum* has expanded to systemic health, with mounting evidence linking it to preterm birth, inflammatory bowel disease, and colorectal cancer [1, 3].

Hung Ton-That (ed.), *Fusobacterium nucleatum: Methods and Protocols*, Methods in Molecular Biology, vol. 3055,
https://doi.org/10.1007/978-1-0716-5388-3_5,

F. nucleatum is taxonomically diverse, comprising at least four well-defined subspecies: *nucleatum* (FNN), *polymorphum* (FNP), *animalis* (FNA), and *vincentii* (FNV) [4, 5]. These subspecies exhibit distinct genomic content, metabolic capabilities, and pathogenic potential [6–8]. Among these, FNA has emerged as a clinically relevant subspecies, frequently isolated from tooth accesses [9] and enriched in colorectal tumor tissues. In particular, clade C2 of FNA has been shown to dominate the colorectal cancer niche [10].

Despite its clinical relevance, genetic studies of FNA have been severely constrained by its low transformation efficiency and resistance to conventional genetic tools. Most existing systems were optimized for the genetically tractable FNN strain ATCC 23726 and are largely ineffective in FNA strains [11].

To circumvent these challenges, we recently developed a conditional plasmid system tailored to genetically recalcitrant *F. nucleatum* strains [12]. This system uses an inducible mechanism to regulate the expression of *repA*, the gene essential for replicating a pCWU6-based shuttle plasmid [13]. The system combines a constitutive *fdx* promoter with a theophylline-responsive riboswitch, enabling controlled plasmid replication. When the inducer is present, the plasmid replicates normally; upon inducer withdrawal, replication ceases, and homologous recombination between the plasmid and the chromosome is favored under antibiotic selection. Our earlier implementation used a *sacB*-based counterselection system, but high sucrose concentrations necessary for effective counterselection impeded bacterial growth [12]. To overcome this, we replaced *sacB* with *mazF*, a toxin gene under the control of an anhydrotetracycline (ATc)-inducible promoter. Upon exposure to ATc, only cells that have lost the plasmid via a second recombination event can survive, significantly improving selection efficiency. In this chapter, we present a comprehensive protocol for applying this optimized system to generate markerless gene deletions in FNA. As a demonstration, we detail the deletion of *luxS* in strain 7_1, a human gut-derived isolate [14]. *LuxS* encodes an essential enzyme in AI-2 production, a signaling molecule involved in bacterial quorum sensing, interspecies communication, and virulence regulation [15].

2 Materials

Unless otherwise noted, all media and solutions should be prepared with deionized Milli-Q water and stored at room temperature (*see* **Note 1**).

2.1 Preparation of Electrocompetent F. nucleatum subsp. animalis Strain 7_1

1. Anaerobic chamber set at 37 °C, equilibrated with an atmosphere of 94% N_2, 3% H_2, and 3% CO_2.
2. Hungate anaerobic tube (16 × 125 mm; Chem Glass Life Science).
3. Butyl rubber stopper and screw cap (9 mm opening).
4. 1% and 10% cysteine solutions: Dissolve 1 g or 10 g of L-cysteine hydrochloride in 100 mL of distilled water. Place in an anaerobic chamber for 10 min to deoxygenate. Dispense 10 mL of aliquots into Hungate tubes, seal, and autoclave at 121 °C for 20 min. Store at room temperature for up to 2 years.
5. 10-cm Petri dishes.
6. 10- and 25-mL serological pipettes.
7. 1-mL disposable syringe with needle.
8. 16 × 150-mm glass culture tubes.
9. 16-mm culture tube caps.
10. TSP broth: In a 500-mL bottle, mix 400 mL of water, 15 g of tryptic soy broth, and 5 g of Bacto™ peptone. Stir until dissolved, adjust to 500 mL with water, and autoclave at 121 °C for 20 min.
11. Working TSPC broth: Use a 10-mL serological pipette to transfer 8 mL of sterile TSP broth into each 16 × 150-mm glass culture tube. Then, using a 1-mL syringe, add 0.4 mL of a sterile 1% L-cysteine hydrochloride solution directly to the medium. Shaking or mixing the tubes is not necessary. Once the cysteine is added, the broth is ready to culture fusobacteria (*see* **Note 2**).
12. TSPC agar plates: Add 400 mL of distilled water to a 1-L glass bottle and dissolve 15 g of tryptic soy broth, 5 g of Bacto™ peptone, and 7.5 g of agar. Adjust the final volume to 500 mL with additional water, and then autoclave the mixture at 121 °C for 15 min. After sterilization, cool the medium to 55 °C in a water bath. Add 5 mL of a 10% cysteine solution and mix thoroughly by gently swirling the bottle by hand. Pour approximately 24 mL of the agar medium into each 10-cm Petri dish. Once the agar has solidified, invert the plates and dry them in a 37 °C incubator. Plates should be used on the same day.
13. Metal inoculating loops.
14. Ice-cold 10% glycerol.
15. Sterile water.
16. 50-mL screw cap tubes.

2.2 Construction of Gene Deletion Plasmid pBCG10Δ luxS

1. Chloramphenicol stock (30 mg/mL): Dissolve 0.3 g in 10 mL of 95% ethanol. Sterilize using a 0.2-μm syringe filter.
2. Competent *E. coli* DH5α cells.
3. Water bath set to 42 °C.
4. LB broth.
5. LB agar plates with 20 μg/mL of chloramphenicol.
6. 1.5 mL-microcentrifuge tubes.
7. PCR supplies: High-fidelity polymerase (e.g., 2× RepliQa HiFi ToughMix), sterilized water, PCR machine, and PCR tubes (*see* **Note 3**).
8. Custom primers:

   P1: CGTTAAAAAGACTTTAGAAGAATTAAAAGCGGAAC TTTGGTTGGAAAAACTG.

   P2: TGCCACATTCCACAGGATTAACTTGAAAACTTGCT ATTTTCTC.

   P3: GTTTTCAAGTTAATCCTGTGGAATGTGGCAATTAT TTAG.

   P4: TTACTTTTTTCAATTTTTTCTAATTCAATCGTGCCA AAGTAAATACATCATCC.

   P5: CTATTTGGAATGATAGCAGCAG.

   P6: CTACTGCGACGTCCATTATTG.

   BCG10F: GCTTTTAATTCTTCTAAAGTCTTTTTAAC.

   BCG10R: GATTGAATTAGAAAAAATTGAAAAAAG.
9. Agarose gel electrophoresis supplies: 50× TAE buffer, 1.5% (w/v) agarose in TAE buffer, ethidium bromide for visualization, 1 kb DNA standard ladder, agarose electrophoresis chamber systems, and appropriate power supply.
10. *F. nucleatum* 7_1 overnight culture in TSPC as PCR template.
11. Conditional plasmid pBCG10 (Addgene ID: 232323). *repA* is regulated by the *pfdx*-riboswitch E, while the toxic gene *mazF* is controlled by a tetracycline-inducible system (TetR). The *tetR-mazF* cassette, derived from plasmid pVoPo-04 [16], replaced the *sacB* gene in pZP06C (*see* Fig. 1) [12].
12. Gel extraction kit.
13. Nanophotometer N60 (IMPLEN) for DNA concentration measurement.
14. Gibson assembly master mix kit (New England Biolabs or equivalent).
15. Plasmid DNA minipreps purification kit.
16. APEX 2× Taq RED Master Mix.
17. SOC medium.

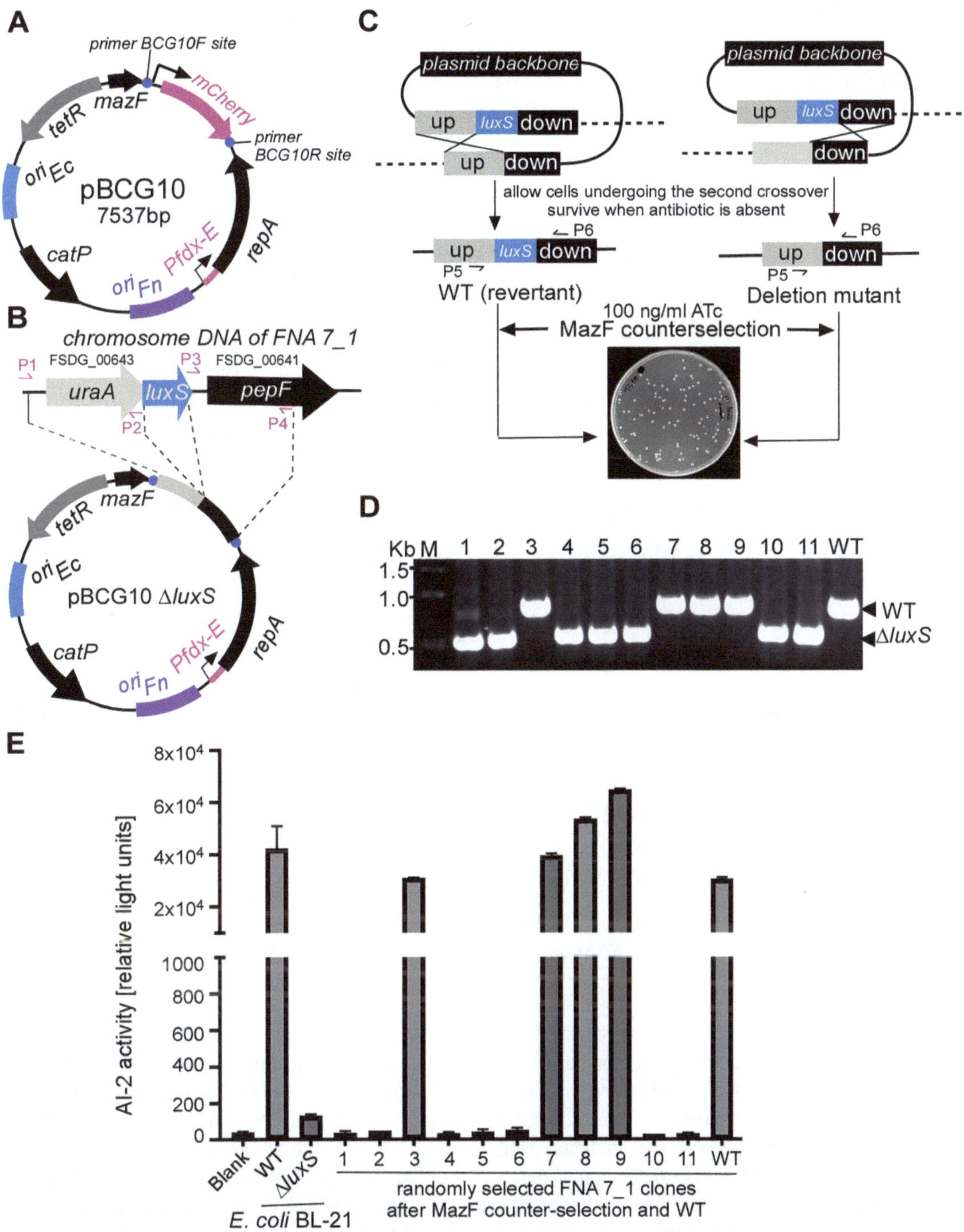

Fig. 1 Construction and validation of a Δ*luxS* mutant in *Fusobacterium nucleatum* subsp. *animalis* strain 7_1. (**a**) Schematic map of the conditional suicide plasmid pBCG10. Plasmid replication in *F. nucleatum* is regulated by a theophylline-inducible system: *repA* expression is driven by a constitutive *fdx* promoter and modulated by a theophylline-responsive riboswitch (Riboswitch E). The toxin gene *mazF*, used for counterselection, is under the control of an anhydrotetracycline (ATc)-inducible promoter (*tetR* system). The plasmid also includes *oriFn* and *repA* for replication in *F. nucleatum*, *oriEc* for propagation in *E. coli*, and *catP* for chloramphenicol resistance. BCG10F/R sites are indicated. (**b**) Plasmid design for deletion of the *luxS* gene via double homologous recombination. pBCG10 contains ~1.5 kb homologous regions upstream (UP) and downstream

2.3 Transformation of the Deletion Plasmid into FNA 7_1

1. *F. nucleatum* subsp. *animalis* strain 7_1 (gift from Dr. Daniel Slade, Virginia Tech).
2. *E. coli* SZU604: This *E. coli* strain expressing three methyltransferase from the FNN strain ATCC 25586 (gift from Dr. Xue Liu, Shenzhen University) [17].
3. Electrocompetent *E. coli* SZU604.
4. Purified plasmid pBCG10Δ*luxS*.
5. 15- and 50-mL polypropylene centrifuge tubes.
6. Competent FNA strain 7_1 cells.
7. Gene Pulser Xcell modular electroporation systems (Bio-Rad).
8. Pre-chill 0.1-mm Bio-rad gene pulser cuvettes.
9. 40 mM theophylline solution: Dissolve 0.228 g in 40-mL TSP, filter-sterilize—store at RT for up to 1 month.
10. 5 mg/mL thiamphenicol stock.
11. 100 μg/mL of anhydrotetracycline (ATc) stock.
12. Parafilm.
13. TSPC agar plates with 5 μg/mL thiamphenicol and 2 mM theophylline.
14. 10% glycerol (autoclaved).

2.4 Generation and Screening of the luxS *Mutant of FNA 7_1*

1. TSPC broth with 5 μg/mL thiamphenicol and 2 mM theophylline.
2. TSPC broth with 5 μg/mL thiamphenicol and 0 mM theophylline.
3. TSPC agar plates with 5 μg/mL thiamphenicol and 0 mM theophylline.
4. TSPC broth with 0 μg/mL thiamphenicol.
5. TSPC agar plates with 100 ng/mL ATc.

Fig. 1 (continued) (DOWN) of the *luxS* coding sequence. The genomic organization of the *luxS* locus in strain 7_1 is shown, with *uraA* and *pepF* flanking *luxS*. (**c**) Gene deletion strategy. The first homologous recombination event integrates the plasmid into the chromosome. Subsequent growth without antibiotic (thiamphenicol) or theophylline allows a second recombination event to resolve the integrated plasmid. This resolution can yield either a clean in-frame Δ*luxS* deletion or restoration of the wild-type allele. Counterselection using ATc induces *mazF* expression, eliminating plasmid-retaining cells and enriching for true recombinants. (**d**) PCR screening of candidate Δ*luxS* colonies. Colony PCR using primer pair P11/P12 was performed on 12 randomly selected colonies. Bands of ~530 bp indicate successful *luxS* deletion, while bands of ~860 bp correspond to wild-type alleles. Seven of the 12 colonies were confirmed as Δ*luxS* mutants. (**e**) Functional validation of the Δ*luxS* mutant using an AI-2 bioassay. Cell-free supernatants from wild-type FNA 7_1 induced strong luminescence in the *Vibrio harveyi* BB170 reporter strain, comparable to the positive control (*E. coli* BL21 WT). In contrast, supernatants from Δ*luxS* mutants showed markedly reduced luminescence, confirming loss of AI-2 production

6. 1.5-mL cryogenic tubes.
7. 60% glycerol (autoclaved).

2.5 AI-2 Assay of FNA 7_1 ΔluxS Mutant

1. 0.22-μm filter (Millipore, Bedford, MA, USA).
2. *Vibrio harveyi* BB170 reporter strain.
3. *E. coli* BL-21 WT and Δ*luxS* control strains.
4. Autoinducer bioassay medium: 0.3 M of NaCl, 0.05 M of $MgSO_4$, 0.2% casamino acids, 10 μM of KH_2PO_4, 1 μM of L-arginine, 20% glycerol, 0.01 μg/mL of riboflavin, and 1 μg/mL of thiamine.
5. Black flat-bottomed 96-well microplate (Greiner Bio-one #655096).
6. GloMax® Navigator Microplate Luminometer.

3 Methods

3.1 Preparation of Electrocompetent F. nucleatum subsp. animalis Strain 7_1

1. Streak FNA 7_1 from a frozen stock onto a TSPC agar plate, and incubate anaerobically at 37 °C for 3 days.
2. Inoculate 2–3 colonies into 8-mL TSPC broth, and incubate statically overnight at 37 °C.
3. Transfer the overnight culture into 42-mL fresh TSPC broth in a 50-mL tube, and grow until OD_{600} ~ 1.2.
4. Centrifuge at 7200 g for 3 min at 4 °C. Discard the supernatant.
5. Wash the pellet twice with 40 mL of sterile Milli-Q water.
6. Resuspend in 30 mL of ice-cold 10% glycerol, centrifuge again, and finally resuspend in 1 mL of ice-cold 10% glycerol (*see* **Note 4**).
7. Aliquot 200 μL per tube, snap-freeze in dry ice-ethanol, and store at −80 °C.

3.2 Construction of Gene Deletion Plasmid pBCG10Δ luxS

1. Inoculate *E. coli* carrying plasmid pBCG10 into 6 mL of LB broth and incubate overnight at 37 °C.
2. Isolate the pBCG10 plasmid from *E. coli* using a plasmid DNA preparation kit.
3. Use the 2× RepliQa HiFi ToughMix (Quantabio, Cat. #: 95200) to amplify the ~6.3 kb of linear pBCG10 plasmid backbone. Set up two 50 μL of PCR reactions using primers BCG10F and BCG10R and 2 μL of overnight culture as template (*see* **Note 5**).

PCR Setup (50 μL):

25 μL of 2× repliQa HiFi ToughMix.

2 μL of BCG10F.

2 μL of BCG10R.

2 μL of *E. coli* pBCG10 culture.

19 μL of Nuclease-free water

4. Identify *luxS* in the genome of Strain 7_1 using BioCyc annotation (Gene ID: FSDG_00642), flanked by *uraA* and *pepF* (*see* Fig. 1b).
5. Use primers P1/P2 to amplify 1.5 kb upstream (UP) of *luxS* and P3/P4 to amplify 1.5 kb downstream (DOWN) region. P2 and P3 contain 20 bp overlaps for Gibson assembly. P1 and P4 contain 29- and 27-nt overlaps with the pBCG10 backbone.
6. Combine the UP and DOWN fragments with the linearized pBCG10 backbone in a 20-μL Gibson Assembly reaction:

X μL pBCG10 backbone (100 ng).

Y μL UP fragment (100 ~ 200 ng).

Z μL DOWN fragment (100 ~ 200 ng).

10 μL 2× Gibson Assembly Master Mix.

(10– X—Y—Z) μL Deionized H_2O.

Incubate at 50 °C for 15 min. Store on ice or at −20 °C.

7. Transformation into *E. coli:*
   - (i) Thaw NEB 5-alpha or DH5α competent cells on ice (~10 min).
   - (ii) Add 5 μL of the assembly reaction to 50 μL of competent cells. Mix gently.
   - (iii) Incubate on ice for 15–40 min.
   - (iv) Heat shock at 42 °C for 50 s, and then return to ice for 2 min.
   - (v) Add 950 μL SOC medium, transfer to a culture tube, and incubate at 37 °C for 1 h with shaking.
   - (vi) Plate 100–200 μL on LB agar with 15 μg/mL chloramphenicol.
   - (vii) Incubate plates overnight at 37 °C.
8. Select (# of colonies) colonies for screening, and perform colony PCR with primers P5/P6.

Colony PCR (25 μL total):

12.5 μL of APEX 2× Taq RED Master Mix

1.25 μL of Primer P5

1.25 μL of Primer P6.

10 μL of nuclease-free water.

Analyze 8 μL of PCR product on agarose gel to verify the amplicon.

9. Extract and purify plasmids using a miniprep kit, measure the concentrations with a NanoDrop, and send samples for whole plasmid sequencing. Store sequence-confirmed plasmids at −20 °C (*see* **Note 6**).

3.3 Transformation of pBCG10ΔluxS (FNA 7_1) in E. coli SZU604

1. Preparation of electrocompetent SZU604: Grow *E. coli* SZU604 in LB broth to mid-log phase in a 50-mL conical tube. Harvest the cells, and prepare them to follow the electrocompetent FNA strain 7_1 (*see* Subheading 3.1).
2. Electroporate pBCG10Δ*luxS* into SZU604:
   (i) Pre-chill 0.1 cm of Bio-Rad Gene Pulser electroporation cuvettes at −20 °C.
   (ii) Thaw prepared *E. coli* SZU604 competent cells on ice (~10 min).
   (iii) Aliquot 1 mL of SOC medium into sterile 1.5-mL microcentrifuge tubes.
   (iv) Add 5 μL (~100 ng) of pBCG10Δ*luxS* plasmid DNA (from Subheading 3.3) to 100 μL of thawed competent cells, and gently mix. Incubate on ice for 10 min.
   (v) Transfer the mixture into a chilled cuvette.
   (vi) Perform electroporation using the following settings: 2.5 kV, 25 μF, and 200 Ω (Gene Pulser Xcell system, Bio-Rad).
   (vii) Immediately transfer the electroporated cells to 1 mL of SOC medium.
   (viii) Incubate at 37 °C for 1 h under aerobic conditions.
   (ix) Plate 100–200 μL of the culture on LB agar supplemented with 15 μg/mL of chloramphenicol.
   (x) Incubate the plates at 37 °C for 24 h under aerobic condition.
3. Select three colonies and inoculate into 6 mL LB broth with 15 μg/mL chloramphenicol. Incubate overnight at 37 °C with shaking under aerobic conditions.
4. Purify plasmid DNA using a miniprep kit and determine DNA.

3.4 Transformation of pBCG10ΔluxS into FNA 7_1 and Mutant Selection

1. Thaw electrocompetent FNA 7_1 cells on ice for ~10 min.
2. Prepare 1 mL of TSPC + theophylline medium by mixing: 9 mL of TSP medium, 0.5 mL of 1% L-cysteine, and 0.5 mL of 40 mM theophylline. Dispense 1 mL per 1.5-mL microcentrifuge tube.

3. Mix 10 μL (~1 μg) of methylated pBCG10Δ*luxS* plasmid DNA with 100 μL of thawed FNA 7_1 competent cells. Incubate on ice for 10 min.
4. Transfer the mixture into a pre-chilled 0.1-cm electroporation cuvette.
5. Electroporate using Bio-Rad Gene Pulser Xcell at 2.5 kV, 25 μF, and 200 Ω (*see* **Note 7**).
6. Immediately add the entire contents to 1 mL of TSPC + theophylline medium.
7. Incubate cells anaerobically at 37 °C for 3 h.
8. Plate 100–200 μL onto TSPC agar plates containing 5 μg/mL thiamphenicol and 2 mM theophylline.
9. Incubate plates anaerobically at 37 °C for 3 days.
10. Pick three thiamphenicol-resistant colonies, and inoculate into 8 mL TSPC broth containing 5 μg/mL thiamphenicol and 2 mM theophylline. Incubate overnight anaerobically at 37 °C.
11. Subculture 8 μL of the overnight culture into 8 mL TSPC broth containing 5 μg/mL thiamphenicol (no theophylline). Incubate anaerobically overnight. This step promotes plasmid integration into the chromosome via homologous recombination.
12. Dilute the overnight culture 1000 and 10,000 times using fresh TSPC medium. Take 100 μL from the two diluted cultures, and spread them on TSPC agar plates containing 5 μg/mL thiamphenicol. Then, incubate the plates under anaerobic conditions for 3 days at 37 °C.
13. Pick up three colonies resistant to thiamphenicol. Inoculate these colonies into an 8-mL TSPC medium containing 5 μg/mL thiamphenicol. Incubate under anaerobic conditions overnight at 37 °C.
14. Transfer 8 μL of overnight culture to fresh 8 mL TSPC broth, and incubate overnight under anaerobic conditions at 37 °C (to allow the cells to undergo the second crossover event, survival, and propagate. The second crossover event results in the loss of the integrated plasmid, *see* Fig. 1c).
15. Dilute the overnight culture 1000 and 10,000 times using fresh TSPC medium. Take 100 μL from the two diluted cultures, and spread them on TSPC agar plates containing 100 ng/mL anhydrotetracycline (*see* Fig. 1c). Then, incubate the plates under anaerobic conditions for 3 days at 37 °C. This allows for counterselection due to the toxic expression of the toxin mazF, and only bacteria that have lost the plasmid can grow.

16. Use sterile polyester-tipped swabs to pick 10–20 single anhydrotetracycline-resistant colonies from **step** 7, and streak them onto TSPC plates and TSPC plates containing 5 μg/mL thiamphenicol to test thiamphenicol sensitivity. Incubate for 3 days under anaerobic conditions at 37 °C.
17. By colony PCR, select 11 thiamphenicol-sensitive strains to screen for mutants lacking *luxS* using primers P5/P6 (*see* Fig. 1c). The results revealed that 7 of the 11 randomly chosen colonies were deletion mutants, while the other 4 reverted to the wild type during the second recombination event. Bands of approximately 530 bp indicate *luxS* deletion, whereas bands of about 860 bp indicate a wild-type (WT) genotype (*see* Fig. 1d) (*see* **Note 8**).
18. The Δ*luxS* clones are prepared for storage by inoculating individual colonies into 8 mL TSPC overnight under anaerobic conditions at 37 °C.
19. Prepare glycerol stock of Δ*luxS* strain for −80 °C storage. Grow individual mutant colonies in 8 mL TSPC medium in 16-mm culture tubes with caps. Transfer the culture to a 15-mL tube and spin at 4500 g for 5 min.
20. Discard the supernatants and resuspend the cell pellets in a final volume of 600 μL of TSPC medium.
21. Transfer the resuspended cultures to appropriately labeled cryogenic tubes. Add 400 μL of 50% glycerol and gently mix.
22. Store the cryogenic tubes in the −80 °C freezer.

3.4.1 AI-2 Assay

1. Grow FNA wild-type and Δ*luxS* mutant strains in 8 mL TSPC broth anaerobically at 37 °C until OD_{600} ~ 0.7.
2. Centrifuge cultures at 12,000 × g for 5 min to pellet the cells.
3. Collect the supernatants and filter through a 0.22-μm syringe filter to obtain sterile cell-free supernatants (CS).
4. Prepare the *Vibrio harveyi* BB170 reporter strain by diluting an overnight culture 1:5000 in fresh autoinducer bioassay medium (composition: 0.3 M NaCl, 0.05 M $MgSO_4$, 0.2% Casamino Acids, 10 μM KH_2PO_4, 1 μM L-arginine, 20% glycerol, 0.01 μg/mL riboflavin, and 1 μg/mL thiamine).
5. In sterile 16-mm culture tubes, mix 4.5 mL of the diluted *V. harveyi* culture with 0.5 mL of each FNA-derived CS sample.
6. Incubate the mixtures at 30 °C with shaking for 4 h.
7. Transfer 150 μL of each mixture to individual wells of a black 96-well flat-bottomed microplate.
8. Measure bioluminescence using a GloMax® Navigator Microplate Luminometer at 30-min intervals for 4 h.

9. Include controls: Cell-free supernatant from *E. coli* BL-21 is the positive control, while its *luxS* mutant is the negative control.
10. Compare luminescence outputs to evaluate AI-2 production. The Δ*luxS* mutant is expected to show significantly reduced or absent AI-2 activity relative to wild-type FNA.

4 Notes

1. Pre-reduce all media in an anaerobic chamber for ≥24 h. Prepare agar plates fresh. Maintain strict anaerobic conditions throughout all steps, minimizing oxygen exposure.
2. Always prepare TSPC medium fresh just before use.
3. Use high-fidelity polymerases (e.g., RepliQa HiFi ToughMix, Pfu, Phusion, PrimeSTAR, and AccuPrime) for cloning to avoid errors introduced by Taq.
4. Anaerobic bacteria are sensitive to oxygen. Perform washes/resuspensions in freshly prepared, deoxygenated water and 10% glycerol. Minimize time outside the chamber.
5. Use fresh overnight *E. coli* cultures or resuspend a chunk of −80 °C frozen stock in 100-μL water.
6. Confirm transformants using genotyping primers. For Gibson or substitution constructs, sequence the full plasmid (e.g., via nanopore). Plasmids purified from *E. coli* DH5α cannot be directly transformed into the *F. nucleatum* strain 7_1. Prior to transformation, the plasmid must be introduced into *E. coli* SZU604, which expresses three DNA methyltransferases (MTases) from *F. nucleatum* ATCC 25586. These MTases modify the plasmid DNA to mimic native methylation patterns. Only properly methylated plasmids can be successfully transformed into the *F. nucleatum* 7_1 wild-type strain.
7. *Fusobacterium*, especially non-model strains, transform poorly. Optimize electroporation, cell density, and recovery time.
8. Longer homologous arms improve recombination efficiency. For large gene edits, use extended flanks and design primers for 200–500 bp mutant-specific bands.

Acknowledgments

This work was supported by NIDCR grants DE030895 and DE034542 to C.W.

Conflict of Interest Statement The authors have no conflicts of interest to declare that are relevant to the content of this chapter.

References

1. Han YW (2015) *Fusobacterium nucleatum*: a commensal-turned pathogen. Curr Opin Microbiol 23:141–147. https://doi.org/10.1016/j.mib.2014.11.013
2. Abusleme L, Dupuy AK, Dutzan N, Silva N, Burleson JA, Strausbaugh LD et al (2013) The subgingival microbiome in health and periodontitis and its relationship with community biomass and inflammation. ISME J 7(5): 1016–1025. https://doi.org/10.1038/ismej.2012.174
3. Brennan CA, Garrett WS (2019) *Fusobacterium nucleatum*—symbiont, opportunist and oncobacterium. Nat Rev Microbiol 17(3): 156–166. https://doi.org/10.1038/s41579-018-0129-6
4. Gharbia SE, Shah HN (1992) Fusobacterium nucleatum subsp. fusiforme subsp. nov. and Fusobacterium nucleatum subsp. animalis subsp. nov. as additional subspecies within Fusobacterium nucleatum. Int J Syst Evol Microbiol 42(2):296–298. https://doi.org/10.1099/00207713-42-2-296
5. Kook JK, Park SN, Lim YK, Choi MH, Cho E, Kong SW et al (2013) Fusobacterium nucleatum subsp. fusiforme Gharbia and Shah 1992 is a later synonym of Fusobacterium nucleatum subsp. vincentii Dzink et al. 1990. Curr Microbiol 66(4):414–417. https://doi.org/10.1007/s00284-012-0289-y
6. Ang MY, Dutta A, Wee WY, Dymock D, Paterson IC, Choo SW (2016) Comparative genome analysis of Fusobacterium nucleatum. Genome Biol Evol 8(9):2928–2938. https://doi.org/10.1093/gbe/evw199
7. Ma X, Sun T, Zhou J, Zhi M, Shen S, Wang Y et al (2023) Pangenomic study of *Fusobacterium nucleatum* reveals the distribution of pathogenic genes and functional clusters at the subspecies and strain levels. Microbiology Spectrum:e05184–22. https://doi.org/10.1128/spectrum.05184-22
8. Connolly JP, Kelly L (2025) The physical biogeography of Fusobacterium nucleatum in health and disease. MBio 16(4):e02989–e02924. https://doi.org/10.1128/mbio.02989-24
9. Krieger M, AbdelRahman YM, Choi D, Palmer EA, Yoo A, McGuire S et al (2024) Stratification of *Fusobacterium nucleatum* by local health status in the oral cavity defines its subspecies disease association. Cell Host Microbe 32(4):479–88.e4. https://doi.org/10.1016/j.chom.2024.02.010
10. Zepeda-Rivera M, Minot SS, Bouzek H, Wu H, Blanco-Míguez A, Manghi P et al (2024) A distinct *Fusobacterium nucleatum* clade dominates the colorectal cancer niche. Nature 628(8007):424–432. https://doi.org/10.1038/s41586-024-07182-w
11. Umaña A, Nguyen TT, Sanders BE, Williams KJ, Wozniak B, Slade DJ (2022) Enhanced *Fusobacterium nucleatum* genetics using host DNA methyltransferases to bypass restriction-modification systems. J Bacteriol 204(12): e00279–22. https://doi.org/10.1128/jb.00279-22
12. Zhou P, GC B, Wu C (2025) Development of a conditional plasmid for gene deletion in non-model *Fusobacterium nucleatum* strains. Appl Environ Microbiol:e01816–24. https://doi.org/10.1128/aem.01816-24
13. Wu C, Al Mamun AAM, Luong TT, Hu B, Gu J, Lee JH et al (2018) Forward genetic dissection of biofilm development by *Fusobacterium nucleatum*: novel functions of cell division proteins FtsX and EnvC. MBio 9(2): 10.1128/mbio.00360-18
14. Strauss J, Kaplan GG, Beck PL, Rioux K, Panaccione R, DeVinney R et al (2011) Invasive potential of gut mucosa-derived Fusobacterium nucleatum positively correlates with IBD status of the host. Inflamm Bowel Dis 17(9):1971–1978. https://doi.org/10.1002/ibd.21606
15. Pereira CS, Thompson JA, Xavier KB (2013) AI-2-mediated signalling in bacteria. FEMS Microbiol Rev 37(2):156–181. https://doi.org/10.1111/j.1574-6976.2012.00345.x
16. Ponath F, Zhu Y, Cosi V, Vogel J (2022) Expanding the genetic toolkit helps dissect a global stress response in the early-branching species *Fusobacterium nucleatum*. Proc Natl Acad Sci 119(40):e2201460119. https://doi.org/10.1073/pnas.2201460119
17. Liu L, He Y, Zhang T, Geng R, Hu Y, Luo M et al (2024) Equip *Fusobacterium nucleatum* genetic tool kits with compatible shuttle vectors and engineered intermediatory *E. coli* strains for enhanced transformation efficiency. bioRxiv. https://doi.org/10.1101/2024.07.17.603877

Chapter 6

Gene Silencing by CRISPR Interference in *Fusobacterium nucleatum*

Shiqi Xu, Bibek G C, Kexin Tan, and Chenggang Wu

Abstract

Fusobacterium nucleatum is a strictly anaerobic bacterium associated with periodontal disease and several systemic conditions, including colorectal cancer and adverse pregnancy outcomes. Genetic manipulation in *F. nucleatum* has been limited by poor transformation efficiency and difficulty studying essential genes. To overcome these challenges, we developed a CRISPR interference (CRISPRi) system that enables efficient and reversible gene silencing without altering the genome. This system uses an inducible dCas9 and a customizable sgRNA expressed from a pCWU6-based shuttle plasmid (pZP4C). In this chapter, we present a step-by-step protocol for designing sgRNAs, constructing CRISPRi plasmids, transforming *F. nucleatum* ATCC 23726, and evaluating gene silencing phenotypes. We use the nonessential gene *ftsW*, which encodes a protein required for peptidoglycan synthesis and cell division, as a model target. This protocol is also applicable to other genetically recalcitrant *F. nucleatum* strains, offering a versatile tool for investigating both essential and nonessential gene functions.

Key words *Fusobacterium nucleatum*, CRISPRi, *ftsW*, dCas9, sgRNA, Gene silencing

1 Introduction

Gene inactivation or silencing is a fundamental strategy for elucidating gene function in living organisms. Among the recent technological advances, CRISPR interference (CRISPRi) has emerged as a powerful and efficient method for targeted gene silencing [1]. It has been widely adopted in genetic studies across various bacterial species due to its simplicity and versatility [2, 3]. The CRISPRi system consists of two core components: a catalytically inactive Cas9 protein (dCas9) and a single-guide RNA (sgRNA), which directs dCas9 to the gene of interest. Mechanistically, when the sgRNA binds to the non-template strand of the target gene, the dCas9-sgRNA-DNA complex obstructs RNA polymerase progression, thereby repressing gene transcription (*see* Fig. 1a).

F. nucleatum is a strictly anaerobic, Gram-negative bacterium that predominantly inhabits the oral cavity of humans and animals

Hung Ton-That (ed.), *Fusobacterium nucleatum: Methods and Protocols*, Methods in Molecular Biology, vol. 3055,
https://doi.org/10.1007/978-1-0716-5388-3_6,

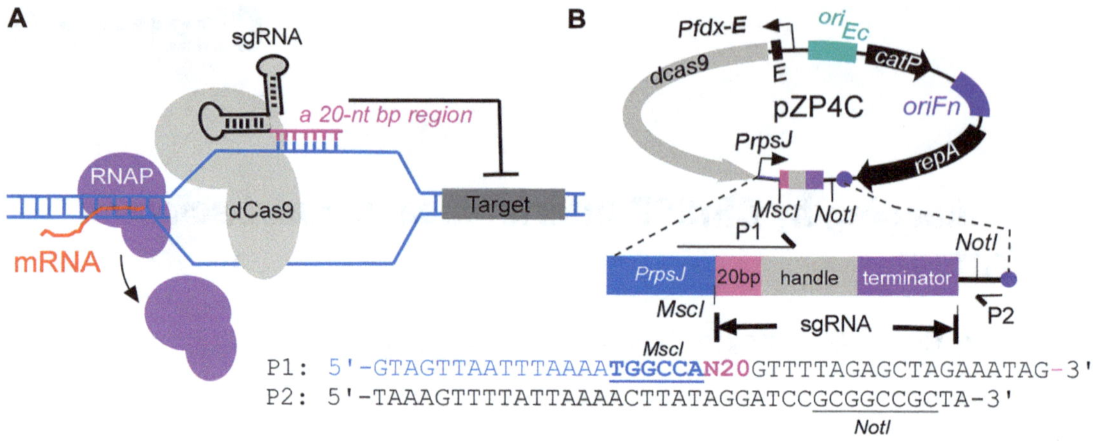

Fig. 1 Schematic overview of CRISPRi-mediated gene silencing in *Fusobacterium nucleatum*. (**a**) Illustration of dCas9-sgRNA complex binding to the non-template strand of the target gene, blocking RNA polymerase and repressing transcription. (**b**) Structure of the CRISPRi plasmid pZP4C, showing riboswitch-inducible expression of dCas9 and constitutive expression of the sgRNA from the *rpsJ* promoter. The sgRNA contains a 20-nt gene-specific targeting region. Primers P1 and P2 are used to generate a new sgRNA cassette by PCR for Gibson assembly

[4]. It is particularly abundant in diseased subgingival biofilms and is considered a key contributor to the onset and progression of periodontitis [5, 6]. In recent years, *F. nucleatum* has also been implicated in various systemic diseases upon translocation to extra-oral sites, including colorectal cancer, adverse pregnancy outcomes, inflammatory bowel disease, and rheumatoid arthritis [6, 7]. Given its clinical relevance, there is increasing interest in uncovering the molecular mechanisms underlying *F. nucleatum* pathogenesis, with gene inactivation as a key approach.

Over the past decade, several genetic tools have been developed for *F. nucleatum*, including single-crossover-based truncation mutation [8], antibiotic cassette replacements [9], and markerless deletion systems utilizing *galK*, *sacB*, or toxin-based counterselection [10–14]. However, these tools have mainly been optimized for ATCC 23726, 12230, and 10953 strains, which exhibit moderate to high plasmid transformation efficiencies. In contrast, many other *F. nucleatum* strains remain genetically recalcitrant, largely due to complex restriction-modification (RM) systems that hinder plasmid uptake [11, 15]. Additionally, traditional gene deletion methods are labor-intensive and unsuited for studying essential genes. To overcome these limitations, we recently developed a CRISPRi system that enables efficient gene silencing and functional studies in *F. nucleatum*, including strains with poor genetic tractability [16]. Unlike gene knockout approaches, CRISPRi does not alter the gene's DNA sequence but suppresses its expression at the transcriptional level, offering a more flexible and efficient route to gene inactivation [1].

In our system, the *dcas9* gene from *Streptococcus pyogenes* is placed under the control of a riboswitch-regulated inducible promoter, while sgRNA expression is driven by the strong *rpsJ* promoter in a pCWU6-based shuttle plasmid named pZP4C (*see* Fig. 1b). Each sgRNA contains three key elements: a 20-nucleotide (nt) target-specific base-pairing region (gRNA), a 42-nt dCas9-binding handle, and a 40-nt transcriptional terminator from *S. pyogenes.* A major advantage of this system is its modularity—targeting a new gene requires only replacing the 20-nt base-pairing segment of the sgRNA, which can be easily achieved via PCR. Specifically, a single primer (P1) enables the construction of customized sgRNAs. P1 contains three components: a 21-nt sequence homologous to the *rpsJ* promoter for Gibson assembly, a central 20-nt customizable region (gRNA) for target gene recognition, and a 19-nt segment complementary to the dCas9 handle. When P1 is paired with primer P2 (the universal reverse primer) for PCR using pZP4C as the template, a 196-bp DNA fragment that includes the new sgRNA is generated. This PCR product can be inserted into MscI/NotI-digested pZP4C using Gibson assembly to create a new CRISPRi plasmid, pZP4C (*new*).

We have successfully applied this system to silence both nonessential and essential genes in *F. nucleatum* [16]. In this chapter, we present a step-by-step protocol for implementing CRISPRi-mediated gene silencing in *F. nucleatum*, using the *ftsW* gene as a model. *ftsW* is an essential gene involved in bacterial cell division, specifically in synthesizing peptidoglycan (PG), a key structural component of the bacterial cell wall [17, 18]. It plays a critical role during cytokinesis, and its silencing leads to marked cellular morphological changes, making it an ideal candidate for demonstrating the utility of the CRISPRi system. Although this protocol is demonstrated using *F. nucleatum* ATCC 23726, it is also applicable to other genetically resistant *F. nucleatum* strains.

2 Materials

2.1 sgRNA Design

1. Genome access for *F. nucleatum* subsp. *nucleatum* ATCC 23726:
   - Available at BioCyc: https://biocyc.org/GCF_003019785/organism-summary.
   - Target gene: *ftsW* (gene ID: C4N14_02545).
2. Online tool for gRNA design: EuPaGDT (Eukaryotic Pathogen CRISPR guide RNA/DNA Design Tool) website: http://grna.ctegd.uga.edu/.

2.2 Construction of CRISPRi Plasmid Targeting ftsW

1. Primer for sgRNA amplification:

   P1(ftsW): GTAGTTAATTTAAAATGGCCAGTTGGAAATA GACTTTCACTGTTTTAGAGCTAGAAATAG.

   P2 (universe reverse primer): AAAGTTTTATTAAAACTTA TAGGATCCGCGGCCGCTA.
2. PCR components and equipment: High-fidelity PCR master mix (e.g., RepliQa HIFI ToughMix, Quantabio), PCR machine, 0.2-mL PCR tubes, ddH2O, and sterile tips.
3. Gel electrophoresis setup: 1× TAE buffer, 1.5% (w/v) agarose in TAE buffer, ethidium bromide for visualization, 1 kb DNA standard ladder, agarose electrophoresis chamber systems, and appropriate power supply.
4. DNA gel purification: Clean razor blade, Gel extraction kit (e.g., QIAquick Gel Extraction Kit), compatible centrifuge, heat block, and 1.5-mL microcentrifuge tubes.
5. pZP4C, a CRISPRi plasmid that is designed for *ftsX* silencing.
6. pZP4C-mCherry, a CRISPRi cloning plasmid (*see* **Note 4**).
7. LB broth.
8. Chloramphenicol stock (30 mg/mL): Dissolve 0.3 g in 10 mL 95% ethanol. Sterilize using a 0.2-μm syringe filter.
9. 37 °C shakers.
10. Plasmid DNA minipreps purification kit.
11. Nanophotometer N60 (IMPLEN) for DNA concentration measurement.
12. Restriction enzymes: MscI and NotI.
13. 37 °C incubator.
14. Gibson assembly master mix kit (New England Biolabs or equivalent).
15. Chemically competent *E. coli* (e.g., Mix & Go cells).
16. LB agar plates with 20 μg/mL of chloramphenicol.
17. A ChemiDoc MP Imaging System (Bio-Rad).
18. Primer for plasmid validation: S-pZP4C-F: GATGAACCAG CAATATCTTATGGAG.

2.3 Transformation of pZP4C(ftsW) into F. nucleatum

1. Bacterial strain: *F. nucleatum subsp. nucleatum* ATCC 23726.
2. Anaerobic chamber: 37 °C with 94% N_2, 3% H_2, and 3% CO_2.
3. Hungate anaerobic tube (16 × 125 mm), butyl rubber stopper, and screw caps.
4. 1% and 10% cysteine solutions: Dissolve 1 g or 10 g of L-cysteine hydrochloride in 100 mL of distilled water. Place in an anaerobic chamber for 30 min to deoxygenate. Dispense

10-mL aliquots into Hungate tubes, seal, and autoclave at 121 °C for 30 min. Store at room temperature for up to 2 years.

5. Glassware and pipettes: 10- and 25-mL serological pipettes, 1-mL disposable syringe with needle, 16 ×150-mm glass culture tubes, and 16-mm culture tube caps.
6. TSP broth: In a 500-mL bottle, mix 400 mL of water, 15 g of tryptic soy broth, and 5 g of Bacto™ peptone. Stir until dissolved, adjust to 500 mL with water, and autoclave at 121 °C for 30 min.
7. TSPC broth: Use a 10-mL serological pipette to transfer 8 mL of sterile TSP medium into each 16 × 150-mm glass culture tube. Then, using a 1-mL syringe, add 0.4 mL of a sterile 1% L-cysteine hydrochloride solution directly to the medium. Shaking or mixing the tubes is not necessary. Once the cysteine is added, the broth is ready to culture the *F. nucleatum*.
8. Materials for making ATCC 23726 competent cells: 50-mL polypropylene centrifuge tubes, a centrifuge compatible with 50 mL tubes, vortex mixer, pre-reduced ddH_2O and 10% glycerol (500 mL each), and dry ice for snap-freezing.
9. Materials for transformation of pZP04C(*ftsW*) into ATCC 23726: Electroporator: Gene Pulser Xcell system (Bio-Rad), pre-chilled 0.1-cm gap cuvettes, thiamphenicol (5 mg/mL stock in 50% alcohol), TSPC agar plates with 5 μg/mL thiamphenicol, ice bucket, and Parafilm for sealing plates (*see* **Note 1**).

2.4 CRISPRi Induction and Phenotypic Assay

1. Induction reagents: 40 mM theophylline solution: Dissolve 0.228 g in 40 mL TSP, filter-sterilize, and store at RT for up to 1 month. TSPC medium with and without 2 mM theophylline.
2. Clean glass slides and coverslips.
3. Micropipettes and sterile filtered tips.
4. Sterilize 1.5-mL Eppendorf tubes for sample handling.
5. Phase-contrast microscope with 1000× oil immersion objective.

3 Methods

3.1 Design Guide RNA Targeting ftsW

1. Retrieve the *ftsW* gene sequence. Visit the Biocyc database (www.biocyc.org).
2. Search for "*Fusobacterium nucleatum* subsp. *nucleatum* ATCC 23736," and navigate to its genome page.

3. In the search bar, enter "C4N14_02545" or "*ftsW*" to locate the gene. Open the gene page, and download the coding sequence in FASTA format (*see* **Note 2**). Download *F. nucleatum* subsp. *nucleatum* ATCC 23736 whole-genome sequence in FASTA format from NCBI page (NCBI Reference Sequence: NZ_CP028109.1).
4. Identify a 20-nt guide sequence targeting *ftsW*. Access the EuPAGDT tool (http://grna.ctegd.uga.edu/).
5. Select "SpCas9: 20nt gRNA, NGG PAM on 3' end." For "Reference Genome," select custom genome and upload *F. nucleatum* subsp. *nucleatum* ATCC 23736 whole-genome sequence in FASTA format.
6. Paste the first half (~500 bp) of the *ftsW* coding sequence into the input box and click "Get guide RNA." Click "gRNA sequence and score" to view the candidate gRNAs.
7. Select an appropriate sgRNA. Choose gRNAs with the following characteristics:
   - High predicted efficiency.
   - Located near the start codon (within the first 100–300 bp).
   - Target the template strand (ensures dCas9 binds the coding strand).
   - Avoid poly-T stretches and predict secondary structure.
   - GC content between 30% and 70%.
   - No off-target.
   - "Potential problems during transcription" column should say "No problem found."
   - Based on these criteria, the gRNA sequence (GTTGGAAA TAGACTTTCACT) was selected for *ftsW* targeting (*see* Fig. 2a).
8. Incorporate the gRNA into the P1 primer and generate primer P1(*ftsW*) (*see* Fig. 2a).

3.2 Construction of CRISPRi Plasmid Targeting ftsW

3.2.1 Day 1: sgRNA Template Amplification

1. Inoculate 6 mL of LB broth containing 20 μg/mL chloramphenicol with *E. coli* carrying plasmid pZP4C.
2. Incubate culture overnight at 37 °C while shaking.
3. PCR amplify the sgRNA region using the 50-μL PCR reaction:
   - 25 μL 2× RepliQa HiFi ToughMix
   - 2 μL overnight culture (as DNA template) (*see* **Note 3**)
   - 2 μL of 10 μM primer P1(*ftsW*)
   - 2 μL of 10 μM universal reverse primer P2
   - 19 μL nuclease-free water.

   And this PCR Program:

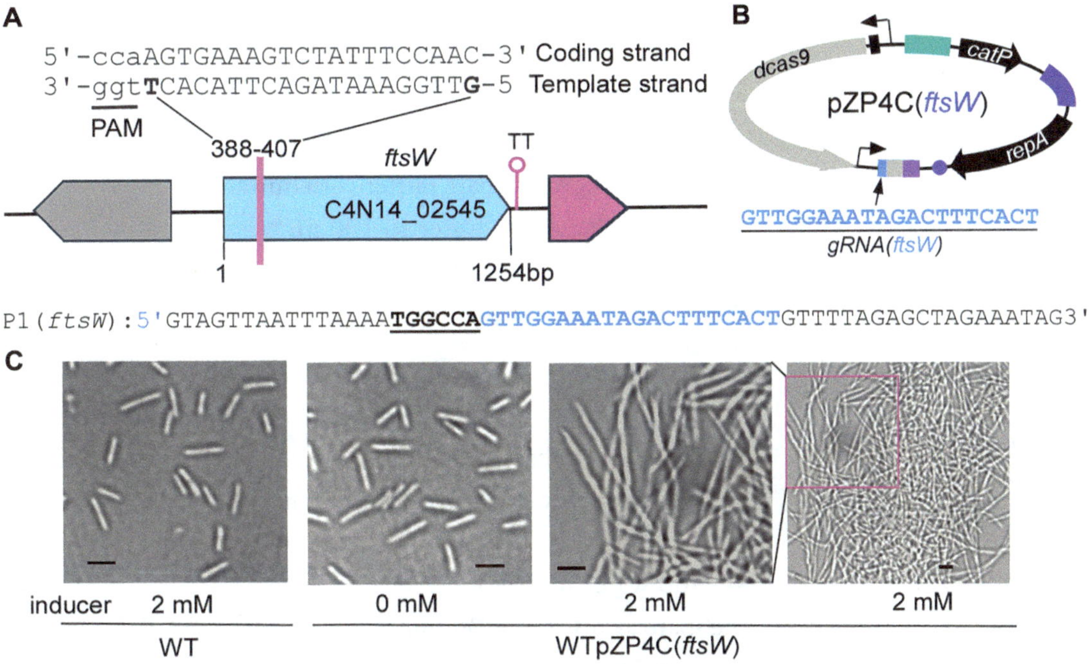

Fig. 2 Design and construction of pZP4C(*ftsW*) for CRISPRi-mediated gene silencing in *Fusobacterium nucleatum*. (**a**) Design of primer P1(*ftsW*), used to generate the sgRNA targeting *ftsW*. The 20-nt gene-specific targeting sequence (gDNA, highlighted in blue) and its position within the *ftsW* coding region are shown. The gene ID in BioCyc is C4N14_02545. "TT" indicates the transcriptional terminator, and "PAM" refers to the protospacer adjacent motif required for Cas9 binding. (**b**) Schematic of the CRISPRi plasmid pZP4C(*ftsW*) carrying the customized sgRNA targeting *ftsW*. The 20-nt targeting sequence is indicated. (**c**) Silencing *ftsW* alters cell morphology in *F. nucleatum*. Phase-contrast microscopy images show wild-type cells and cells carrying pZP4C(*ftsW*), with and without theophylline induction. Cells exhibit filamentous morphology upon CRISPRi induction; scale bar of 2 μm

   (i) Initial denaturation: 98 °C for 6 min.
   (ii) 40 cycles: 98 °C (10 s), 55 °C (6 s), and 68 °C (5 s)
   (iii) Final extension: 68 °C for 2 min.
   (iv) Hold at 4 °C.
4. Run the PCR product on a 1.5% agarose gel.
5. Excise the ~196 bp band and purify using a gel extraction kit.
6. Measure DNA concentration using a Nanophotometer (e.g., IMPLEN N60).

3.2.2 Day 2–3: Preparation of Linearized CRISPRi Plasmid Backbone

1. Grow *E. coli* containing pZP4C-mCherry. Inoculate 6 mL of LB broth (20 μg/mL chloramphenicol) (*see* **Note 4**).
2. Incubate overnight at 37 °C with shaking.
3. Purify plasmid pZP4C-mCherry from 3 mL of the culture using a commercial plasmid miniprep kit.
4. Measure DNA concentration.

5. Digest plasmid using the 100 μL reaction:
   - ~4 μg of pZP4C-mCherry plasmid.
   - 4 μL MscI.
   - 4 μL NotI.
   - 10 μL 10× CutSmart buffer (or equivalent).
   - Add nuclease-free water to 100 μL.
6. Incubate at 37 °C for 3 h.
7. Purify the digested plasmid by gel extraction, removing the *mCherry* insert.

3.2.3 Day 3: Gibson Assembly and Transformation

1. Assemble the final plasmid in a 20-μL reaction:
   - 10 μL 2× Gibson assembly master mix.
   - ≥100 ng digested pZP4C backbone.
   - ≥100 ng purified 196 bp sgRNA PCR product.
   - Add water to 20 μL.
2. Incubate at 50 °C for 30 min.
3. Place on ice.
4. Transform into *E. coli*. Add 5 μL of the Gibson reaction to competent *E. coli* (Mix & Go).
5. Incubate on ice for 8–10 min.
6. Plate the entire mixture onto LB agar plates containing 20 μg/mL chloramphenicol.
7. Incubate 1–2 days at 37 °C.

3.2.4 Day 4–5: Screening and Plasmid Validation

1. Screen recombinant colonies. Pick ten colonies into separate glass culture tubes containing 6 mL of LB + 20 μg/mL chloramphenicol.
2. Grow overnight at 37 °C while shaking.
3. Centrifuge 1 mL from each culture and inspect the pellet:
   (a) Red pellet: plasmid still contains mCherry (nonrecombinant).
   (b) White pellet: likely recombinant plasmid.
4. Select 2–3 white colonies and isolate plasmids using a miniprep kit.
5. Sequence plasmids using primer S-pZP4C-F: GATGAACCAG CAATATCTTATGGAG (*see* **Note 5**).
6. Confirm sgRNA sequence targeting *ftsW* (*see* Fig. 2b).
7. Prepare glycerol stock for downstream use. Mix 800 μL of culture with 200 μL of sterile 50% glycerol.

8. Store at −80 °C for long-term use. The confirmed CRISPRi plasmid targeting *ftsW* is now ready for transformation into *F. nucleatum* ATCC 23726.

3.3 Preparation of F. nucleatum ATCC 23726 Electrocompetent Cells

This section describes the preparation of electrocompetent *F. nucleatum* ATCC 23726 cells for plasmid transformation.

1. Streak ATCC 23736 from a frozen stock onto a TSPC agar plate and incubate anaerobically at 37 °C for 3 days.
2. Inoculate 2–3 colonies into 8 mL TSPC broth, and incubate statically overnight at 37 °C.
3. Transfer the overnight culture into 40 mL of fresh TSPC broth in a 50-mL tube, and grow until OD_{600} ~ 1.2.
4. Centrifuge at 7200 g for 3 min at 4 °C. Discard the supernatant.
5. Wash the pellet twice with 40 mL of sterile Milli-Q water, centrifuging each time under the same conditions.
6. Resuspend in 30 mL of ice-cold 10% glycerol. Centrifuge again and finally resuspend in 1 mL of ice-cold 10% glycerol.
7. Aliquot 100 μL per tube, snap-freeze in a dry ice-ethanol bath, and store at −80 °C for long-term use.

3.4 Transformation of F. nucleatum ATCC 23726 with pZP4C (ftsW)

1. Thaw competent cells. Remove one 100-μL aliquot of electrocompetent *F. nucleatum* ATCC 23726 cells from the −80 °C freezer.
2. Thaw the cells gently by holding the tube in your hand for ~1 min, and then immediately place on ice.
3. Add 10 μL (~100 ng) of pZP4C(*ftsW*) plasmid DNA to the chilled cells. Mix gently by tapping; do not vortex.
4. Incubate on ice for 10 min.
5. Prepare the recovery medium during the incubation: aliquot 1 mL of pre-reduced TSPC broth into a sterile 1.5-mL microcentrifuge tube.
6. Transfer the DNA-cell mixture to a pre-chilled 0.1-cm electroporation cuvette.
7. Electroporate using Bio-Rad Gene Pulser Xcell settings: voltage: 2.5 kV, capacitance: 25 μF, and resistance: 200 Ω.
8. Immediately transfer the electroporated cells into the 1-mL TSPC medium.
9. Incubate the cells anaerobically at 37 °C for 3–5 h to allow recovery.
10. After recovery, divide the culture into two plating conditions for selection:

11. Plate 100 μL directly onto a TSPC agar plate containing 5 μg/mL thiamphenicol.
12. Centrifuge the remaining 1 mL of culture at 7200 g for 3 min. Discard ~900 μL of the supernatant. Resuspend the pellet in the remaining ~100 μL, and plate the full volume on a second selective TSPC agar plate (*see* **Note 6**).
13. Seal plates with Parafilm and incubate anaerobically at 37 °C for 3–4 days.
14. Colonies generally reach ~1 mm in diameter by day 3 and ~ 2 mm by day 4.
15. On day 4, pick three well-isolated colonies for screening, and inoculate each into a separate glass culture tube containing 8 mL TSPC broth with 5 μg/mL thiamphenicol.
16. Incubate statically overnight at 37 °C in the anaerobic chamber (*see* **Note 7**).

3.5 Induction and Assessment of CRISPRi Silencing of ftsW

This section describes how to induce CRISPRi-mediated *ftsW* silencing in *F. nucleatum* ATCC 23726 and evaluate morphological changes.

1. Set up subculture for induction. Inoculate 8 μL of an overnight culture harboring pZP4C(*ftsW*) into 8 mL of pre-reduced TSPC broth with 5 μg/mL thiamphenicol (*see* **Note 8**).
2. Prepare two parallel cultures:
   (a) Without inducer (control).
   (b) With 2 mM theophylline (final concentration).
3. Incubate both cultures anaerobically at 37 °C for 24 h (*see* **Note 9**).
4. After the 24-h induction period, visually compare the turbidity of induced vs. control cultures to assess cell growth and morphology. Since *ftsW* is essential for septation, its silencing is expected to cause cell elongation.
5. Prepare cells for phenotypic assessment. Place 2 μL of each culture (with and without theophylline) onto a clean glass slide.
6. Cover with a coverslip and examine under a phase-contrast microscope at 1000× magnification using oil immersion.
7. Document observations:
   (a) Without inducer: Cells should display typical short, spindle-shaped *F. nucleatum* morphology.
   (b) With inducer: Cells should appear elongated and filamentous, indicating successful silencing of *ftsW*.
8. Confirm CRISPRi effectiveness: the observed filamentous phenotype upon theophylline induction confirms the functionality of the CRISPRi plasmid targeting *ftsW* (*see* Fig. 2c) (*see* **Note 10**).

9. Prepare glycerol stock. Grow 8 mL of WT pZP4C(*ftsW*) culture in TSPC medium supplemented with 5 μg/mL thiamphenicol overnight at 37 °C in an anaerobic chamber.
10. Harvest the cells by centrifugation at 7200 g for 3 min.
11. Discard 7 mL of the supernatant, leaving 1 mL to resuspend the cell pellet.
12. Add 0.5 mL of sterile 60% glycerol to achieve a final glycerol concentration of 20%.
13. Mix thoroughly, transfer the mixture to 2 mL cryovials, and store at −80 °C for long-term preservation.

4 Notes

1. TSPC agar plates placed in the anaerobic chamber can easily dry out due to moisture loss. To minimize evaporation and prevent the plates from drying, it is recommended that the plates be sealed with Parafilm before incubation.
2. Ensure the FASTA file contains a continuous sequence without spaces or line breaks, as CRISPR guide design tools require.
3. Bacterial culture can be used as a DNA template for convenience. The small volume of LB broth typically present does not interfere with the activity of most DNA polymerases. Alternatively, purified plasmid DNA can also be used as the PCR template.
4. For propagation of the pZP4C-mCherry plasmid, *E. coli* XL1-Blue is recommended over DH5α. We observed that plasmid DNA isolated from DH5α often exhibits incomplete digestion with MscI, whereas plasmid DNA prepared from XL1-Blue digests efficiently. This is likely due to methylation of sequences near the MscI recognition site in DH5α, as MscI is known to be sensitive to methylation.
5. Primer S-pZP4C-F anneals to the *rpsJ* promoter region within the pZP4C-Cherry-based plasmid.
6. Typically, transforming ATCC 23726 with 100 ng of a pZP4C-based plasmid yields over 4000 colonies under optimal conditions.
7. *F. nucleatum* cultures typically produce a grayish cell pellet due to hydrogen sulfide (H_2S) production, which reacts with trace metals (e.g., iron) in the medium to form black precipitates such as FeS. Contamination should be suspected if the pellet appears white or any other color. Examine the culture under a microscope to confirm cell morphology and purity.

8. A low inoculation ratio (e.g., 1:1000 dilution) is recommended to enhance the visibility of CRISPRi-mediated gene silencing effects. This is particularly important for genes whose products are required in only minimal amounts to sustain growth. When using a high inoculation density, residual gene products from the overnight culture may support growth, thereby masking the phenotypic consequences of gene silencing.
9. Growth may initially appear severely inhibited when targeting essential genes with CRISPRi. However, sudden outgrowth can occur if incubation extends beyond 2–3 days. The emergence of escape mutants often causes this, typically mutations in the CRISPRi repressor system, which relieve gene silencing or produce some suppressor mutants.
10. To evaluate the effectiveness of CRISPRi silencing, it is recommended to use qRT-PCR to assess changes in transcript levels of the target gene. If the target gene encodes a protein, Western blotting can detect changes in protein expression, provided that a specific antibody is available.

Acknowledgments

This work was supported by NIDCR grants DE030895 and DE034542 to C.W.

Conflict of Interest Statement The authors have no conflicts of interest to declare that are relevant to the content of this chapter.

References

1. Qi LS, Larson MH, Gilbert LA, Doudna JA, Weissman JS, Arkin AP et al (2013) Repurposing CRISPR as an RNA-guided platform for sequence-specific control of gene expression. Cell 152(5):1173–1183. https://doi.org/10.1016/j.cell.2013.02.022
2. Zhang R, Xu W, Shao S, Wang Q (2021) Gene silencing through CRISPR interference in bacteria: current advances and future prospects. Front Microbiol 12:635227. https://doi.org/10.3389/fmicb.2021.635227
3. Call SN, Andrews LB (2022) CRISPR-based approaches for gene regulation in non-model bacteria. Front Genome Edit 4:892304. https://doi.org/10.3389/fgeed.2022.892304
4. Bolstad AI, Jensen HB, Bakken V (1996) Taxonomy, biology, and periodontal aspects of *Fusobacterium nucleatum*. Clin Microbiol Rev 9(1):55–71. https://doi.org/10.1128/CMR.9.1.55
5. Abusleme L, Dupuy AK, Dutzan N, Silva N, Burleson JA, Strausbaugh LD et al (2013) The subgingival microbiome in health and periodontitis and its relationship with community biomass and inflammation. ISME J 7(5):1016–1025. https://doi.org/10.1038/ismej.2012.174
6. Lourenço TGB, Heller D, Silva-Boghossian CM, Cotton SL, Paster BJ, Colombo APV (2014) Microbial signature profiles of periodontally healthy and diseased patients. J Clin Periodontol 41(11):1027–1036. https://doi.org/10.1111/jcpe.12302
7. Brennan CA, Garrett WS (2019) *Fusobacterium nucleatum*—symbiont, opportunist and

oncobacterium. Nat Rev Microbiol 17(3): 156–166. https://doi.org/10.1038/s41579-018-0129-6

8. Kaplan CW, Lux R, Haake SK, Shi W (2009) The *Fusobacterium nucleatum* outer membrane protein RadD is an arginine-inhibitable adhesin required for inter-species adherence and the structured architecture of multispecies biofilm. Mol Microbiol 71(1):35–47. https://doi.org/10.1111/j.1365-2958.2008.06503.x
9. Han YW, Ikegami A, Rajanna C, Kawsar HI, Zhou Y, Li M et al (2005) Identification and characterization of a novel adhesin unique to oral fusobacteria. J Bacteriol 187(15): 5330–5340. https://doi.org/10.1128/JB.187.15.5330-5340.2005
10. Wu C, Al Mamun AAM, Luong TT, Hu B, Gu J, Lee JH et al (2018) Forward genetic dissection of biofilm development by Fusobacterium nucleatum: novel functions of cell division proteins FtsX and EnvC. MBio 9(2). https://doi.org/10.1128/mBio.00360-18
11. Umaña A, Nguyen TT, Sanders BE, Williams KJ, Wozniak B, Slade DJ (2022) Enhanced *Fusobacterium nucleatum* genetics using host DNA methyltransferases to bypass restriction-modification systems. J Bacteriol 204(12): e00279–22. https://doi.org/10.1128/jb.00279-22
12. Ponath F, Zhu Y, Cosi V, Vogel J (2022) Expanding the genetic toolkit helps dissect a global stress response in the early-branching species *Fusobacterium nucleatum*. Proc Natl Acad Sci 119(40):e2201460119. https://doi.org/10.1073/pnas.2201460119
13. Gc B, Zhou P, Wu C (2023) HicA toxin-based Counterselection marker for allelic exchange mutations in *Fusobacterium nucleatum*. Appl Environ Microbiol:e00091–23. https://doi.org/10.1128/aem.00091-23
14. Zhou P, GC B, Hu B, Wu C (2024) Development of SacB-based counterselection for efficient allelic exchange in *Fusobacterium nucleatum*. Microbiol Spectr:e02066–24. https://doi.org/10.1128/spectrum.02066-24
15. Liu L, He Y, Zhang T, Geng R, Hu Y, Luo M et al (2024) Equip *Fusobacterium nucleatum* genetic tool kits with compatible shuttle vectors and engineered intermediatory E. coli strains for enhanced transformation efficiency. bioRxiv:2024.07. 17.603877. https://doi.org/10.1101/2024.07.17.603877
16. Zhou P, GC B, Stolte F, Wu C (2024) Use of CRISPR interference for efficient and rapid gene inactivation in *Fusobacterium nucleatum*. Appl Environ Microbiol 90(2):e01665–23. https://doi.org/10.1128/aem.01665-23
17. Boyle DS, Khattar MM, Addinall SG, Lutkenhaus J, Donachie WD (1997) ftsW is an essential cell-division gene in Escherichia coli. Mol Microbiol 24(6):1263–1273. https://doi.org/10.1046/j.1365-2958.1997.4091773.x
18. Taguchi A, Welsh MA, Marmont LS, Lee W, Sjodt M, Kruse AC et al (2019) FtsW is a peptidoglycan polymerase that is functional only in complex with its cognate penicillin-binding protein. Nat Microbiol 4(4): 587–594. https://doi.org/10.1038/s41564-018-0345-x

Chapter 7

Methods for the Evasion of Restriction-Modification Systems During Genetic Engineering of *Fusobacterium* Species

Martha A. Zepeda-Rivera, Elsa F. McMahon, Kaitlyn N. Lewis, Rutika Gavate, Falk Ponath, and Christopher D. Johnston

Abstract

Genetic manipulation of clinical *Fusobacterium* isolates is severely limited by strain-specific restriction-modification (RM) systems that recognize and degrade foreign DNA. This chapter describes strategies to evade RM defenses and improve transformation efficiencies across diverse *Fusobacterium* lineages. We employ a sequence-based "RM-silencing" method: in silico removal of nonessential regions and introduction of synonymous or GC-preserving nucleotide substitutions to disrupt RM recognition sites within coding and noncoding regions, respectively. Detailed protocols for high-molecular-weight genomic DNA isolation, computational motif identification, plasmid optimization, anaerobic preparation of electrocompetent cells, and optimized electroporation parameters are provided. Application of these methods has enabled successful delivery of replicative plasmids, linear recombination templates, and transposon cassettes into previously intractable *Fusobacterium* clinical isolates. Collectively, these techniques constitute a versatile toolkit to overcome RM barriers and advance functional genetic studies in *Fusobacterium* species.

Key words *Fusobacterium*, Genetic engineering, Restriction-modification systems, SyngenicDNA, Transformation, Electroporation

1 Introduction

Despite the association of *Fusobacterium* species with several disease states [1–4], mechanistic studies to understand their direct roles have been significantly hindered by our inability to genetically engineer clinical isolates of interest. Some success in specific *Fusobacterium* isolates, particularly ATCC model strains, has been achieved, but as with many other microbes of clinical relevance, *Fusobacterium* species overall remain largely genetically intractable, recalcitrant to genetic manipulation using modern in vitro techniques. The most widespread cause of genetic intractability are restriction modification (RM) systems, which enable bacteria to

Hung Ton-That (ed.), *Fusobacterium nucleatum: Methods and Protocols*, Methods in Molecular Biology, vol. 3055,
https://doi.org/10.1007/978-1-0716-5388-3_7,

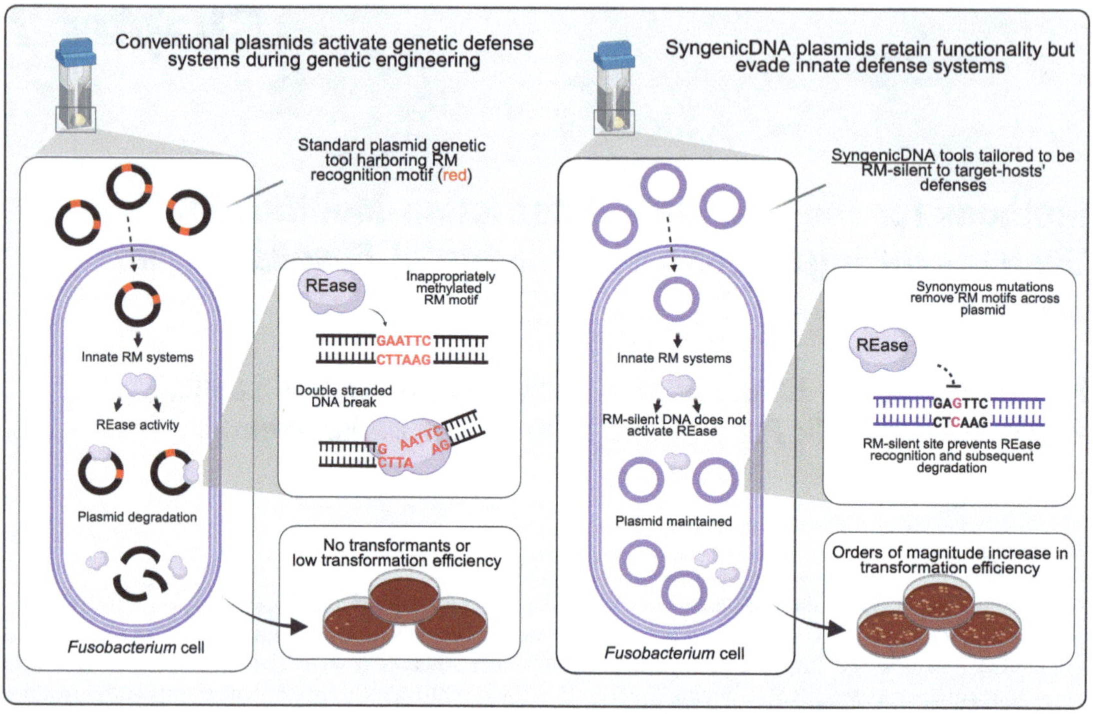

Fig. 1 Schematic depicts (left) transformation with a standard, unmodified plasmid genetic tool as compared to (right) transformation with the SyngenicDNA-tailored version. In the standard genetic tool, red sections depict a highly specific nucleotide recognition motif recognized by a restriction endonuclease (red protein) resulting in double-stranded DNA breaks and subsequent plasmid degradation. Through synonymous mutations, these motifs are removed in the SyngenicDNA-tailored plasmid tool, thereby avoiding restriction endonuclease recognition and subsequent degradation

distinguish self from non-self DNA and selectively degrade the latter. Restriction endonucleases (REase) recognize the methylation status of a highly specific DNA motif and cleave inappropriately methylated DNA, while cognate methyltransferase (MTase) enzymes modify the same target sequence to protect self-DNA from degradation. One approach to overcome RM defenses is to methyl-modify genetic tools to mimic the methylation pattern of the desired host strain [5, 6]. While such mimicry-by-methylation approaches have been successfully applied in *Fusobacterium* [7], their labor-intensive nature coupled with the hypervariability in RM defenses across *Fusobacterium* species, subspecies, and strains has limited their application.

An alternative strain-tailored approach is to eliminate RM-target motifs that exist within the nucleic acid sequence of genetic tools, either through removal of superfluous regions or the introduction of synonymous mutations to achieve RM-silencing [8] (*see* Fig. 1). Here we present a series of methods to apply RM-silent approaches in *Fusobacterium*, ultimately to enhance transformation efficiencies in previously intractable strains.

Through PacBio Single-Molecule Real-Time (SMRT) sequencing [9], the target sequences of a bacterial strain's methyltransferases (methyl-modified nucleotide motifs) are detected at single base resolution. Analysis of genome and methylome data through the restriction enzyme database (REBASE) [10] facilitates identification of RM-associated methyl-modified motifs. Subsequently, through removal of superfluous vector regions and in silico editing of RM-target motifs, genetic tools are sequence-optimized to be RM-silent, invisible to the host's REase, and can be synthesized de novo. Additional protocols for the preparation of *Fusobacterium* electrocompetent cells and subsequent electroporation methods are included here. Through the application of these techniques, we have successfully genetically engineered clinical isolates across multiple *Fusobacterium* lineages using replicative vectors, linearized DNA fragments for homologous recombination, and transposon-based mutagenesis cassettes. These methods are conceptually applicable to any *Fusobacterium* strain of interest.

2 Materials

2.1 Isolation of Fusobacterium High-Molecular-Weight Genomic DNA

1. Filter-sterilized TES buffer: 20 mM of Tris-HCl and 50 mM of EDTA at pH 8.
2. Bioresearch technologies MasterPure Gram-Positive DNA Purification Kit (Item ID MGP04100).
3. Molecular-grade 70% ethanol.
4. Molecular-grade 100% propanol.
5. Sterile water.
6. Sterile 10-μL plastic inoculation loop.
7. Sterile 1× phosphate buffered saline (1× PBS).
8. 1.5-mL and 2-mL Eppendorf tubes.
9. Ice bucket.
10. Dry baths preset to 37 °C and 65 °C.
11. Vortex.
12. Refrigerated centrifuge (for 1.5-mL and 2-mL Eppendorf tubes).

2.2 Identification of RM System Target Motifs in Fusobacterium Strains

1. REBASE: Restriction Enzyme Database (http://rebase.neb.com/) for the identification of restriction-modification systems in *Fusobacterium* genomes.
2. PacBio SMRT Analysis Software (https://www.pacb.com/support/software-downloads/) for detection of methyl-modified motifs at single-base resolution from sequencing data.

3. Geneious, DNAstar, Plasmapper, or equivalent sequence analysis software for alignment, motif identification, and visualization.
4. Data: PacBio SMRT sequencing genome sequence for *Fusobacterium* strain of interest, in the form of a GenBank annotated (.gbk) file.
5. Methylome data (motif_summary.csv file) produced from the SMRT Analysis routine for each *Fusobacterium* strain.

2.3 Optimization of Plasmid Vectors and Application of SyngenicDNA Approaches in Silico

1. Benchling, SnapGene, or equivalent molecular biology software for design, annotation, and in silico mutagenesis of plasmid vectors.
2. In silico tools for determination of codon bias (e.g., Kazusa Codon Usage Database: https://www.kazusa.or.jp/codon/).
3. Codon usage table generated for each *Fusobacterium* of interest based on genome sequence data.

2.4 Preparation and Transformation of Electrocompetent Fusobacterium Cells

1. Sterile cotton swabs.
2. Tryptic soy broth (TSB); commercially available.
3. Yeast extract.
4. L-cysteine hydrochloride.
5. Filter-sterilized supplemented tryptic soy broth (TSB); TSB, 2.5% yeast extract, and 0.4 mg/mL of L-cysteine hydrochloride.
6. 250-mL sterile glass bottle.
7. 50-mL sterile conical tubes.
8. 1.5-mL and 5-mL Eppendorf tubes.
9. Refrigerated centrifuge (for 5-mL Eppendorf tubes and 50-mL conical tubes).
10. Sterile 100% glycerol.
11. Sterile water.
12. Ice buckets.
13. Electroporation cuvettes (0.1-cm gap).
14. Electroporator.
15. Antibiotic according to plasmid selection marker.

3 Methods

3.1 Isolation of *Fusobacterium* High-Molecular-Weight Genomic DNA for PacBio Sequencing

To be performed inside ventilated hood

1. Plate *Fusobacterium* strain on agar plate (*see* **Note 1**), and incubate anaerobically at 37 °C (*see* **Note 2**) for 24–72 h.
2. Using sterile inoculation loop, scrape up cells from a plate, and resuspend in 1.5 mL of PBS in a 2-mL Eppendorf tube.
3. Centrifuge samples at approximately 2300 rcf for 10 min.
4. Pipette out the supernatant.
5. Wash pellets by resuspending in 1.5 mL of TES buffer.
6. Add 300 μL of TE buffer from kit and pipette up and down gently to resuspend.
7. Add 1 μL of lysozyme from the MasterPure kit to each tube. Flick and invert to ensure mixing.
8. Pulse spin to draw the liquid down and incubate in a 37 °C dry bath for 20 min (*see* **Note 3**).
9. Add 300 μL of Gram-positive lysis solution from the kit and briefly vortex.
10. Add 2 μL of proteinase K from the kit, invert to mix well, pulse spin, and incubate in 65 °C dry bath for 15 min.
11. Transfer to a 37 °C dry bath and let cool for 5 min.
12. Place tubes on ice for 3–5 min.
13. Keeping tubes on ice as much as possible, add 350 μL of MPC protein precipitation reagent from the kit, and pulse vortex 5–10 s. Place tubes back on ice.
14. Centrifuge samples at 4 °C and at approximately 14,000 rcf for 10 min.
15. Using a 1-mL pipette, carefully and slowly transfer the supernatant into a clean 1.5-mL Eppendorf tube, avoiding any sediment.
16. Centrifuge samples at 4 °C and at approximately 14,000 rcf for 2 min.
17. Using a 1-mL pipette, carefully and slowly transfer the supernatant into a clean 1.5-mL Eppendorf tube.
18. Add 2 μL of RNAse A from the kit, and mix thoroughly by inverting and flicking the tube manually. Pulse spin to bring the liquid down.
19. Incubate in a 37 °C dry bath for 30 min.
20. During incubation, prepare one 2-mL Eppendorf tube per sample. To each tube, add 700 μL of molecular-grade 100% isopropanol. Place tubes on ice.

21. During incubation, prepare 2 mL per sample of 70% ethanol, using molecular-grade ethanol and sterile water. Place total volume in a conical tube and place conical tube on ice.
22. Using a 1-mL pipette, carefully and slowly transfer supernatant into a 2-mL Eppendorf tube pre-filled with chilled isopropanol. Quickly close the cap, and invert the tube roughly up to 30–40 times until the formation of DNA spindles is observed (*see* **Note 4**). Place each sample back on ice until all samples are ready to proceed.
23. Carefully loop the DNA spindles around a sterile p1000 pipette tip, and immediately place into a 2-mL tube containing pre-chilled 70% ethanol (*see* **Note 5**).
24. Gently invert the tube to wash the pellet.
25. Centrifuge samples at 4 °C and at approximately 10,000 rcf for 2 min.
26. Using a 1-mL pipette, carefully remove ethanol.
27. Add 1 mL of 70% pre-chilled ethanol to dislodge and wash pellet.
28. Centrifuge samples at 4 °C and at approximately 10,000 rcf for 2 minutes.
29. Using a 1-mL pipette, carefully remove ethanol.
30. Using a 200-μL pipette, remove as much residual ethanol as possible.
31. Air dry the pellet in the hood, leaving the tube uncapped for at least 30 minutes.
32. During **step 30**, pre-heat sterile water in a 1.5-mL Eppendorf tube in 37 °C dry bath. You will require a total of 50–200 μL per DNA pellet.
33. Depending on the size of the pellet, add 50–200 μL of pre-warmed sterile water to pellet (*see* **Note 6**).
34. Incubate tube (cap closed) in a 37 °C dry bath until pellet dissolves (can take up to 30 minutes).
35. To quality check each sample, measure the 260/280 and 260/230 values using a Nanodrop. For a more accurate concentration measurement, measure the concentration using a Qubit. To ensure that DNA is minimally sheared for the use of PacBio sequencing, run 500 ng of sample on a 1% agarose gel (*see* **Note 7**).
36. PacBio-sequence each DNA sample, ensuring with the core facility that enzyme kinetics are switched on. Enzyme kinetics are used to deduce methyl-modified nucleotide motifs, which is crucial for downstream optimization of plasmid vectors to increase transformation success.

3.2 Identification of RM System Target Motifs in Fusobacterium Strains

1. Obtain PacBio sequencing data for *Fusobacterium* strain, ensuring enzyme kinetics are enabled to detect methyl-modified nucleotide motifs.
2. Submit genome file (.gbk) and modified motifs (.csv file, obtained from PacBio BaseMod analysis) to the REBASE database (http://rebase.neb.com).
3. Use the REBASE output reports to identify modified motifs associated with corresponding methyltransferases within the *Fusobacterium* genome (*see* **Note 8**).

3.3 Optimization of Plasmid Vectors and Application of SyngenicDNA Approaches in Silico

1. Annotate plasmid or genetic tool sequences to determine the precise location of open reading frames (ORFs), promoters, terminators, and other critical features (*see* **Note 9**).
2. Using Benchling or SnapGene, screen for the presence of RM-associated motifs from each strain of interest within each genetic tool sequence (*see* **Note 10**).
3. Annotate each of the identified RM sequence motifs as a new feature "RM motif" (*see* **Note 11**).
4. For RM motifs located within coding regions, introduce a single synonymous mutation to alter one or more nucleotides within the motif in a way that eliminates the RM recognition site, while adhering to the codon usage table for the *Fusobacterium* strain of interest. The mutation should be nonredundant, ensuring that it unambiguously disrupts the RM motif (*see* **Notes 12** and **13**).
5. For RM motifs located within noncoding regions, there is not an option of introducing synonymous mutations (as it is noncoding). In these cases, instead introduce nucleotide modifications that preserve the overall GC content of these specific regions by substituting for its complementary nucleotide ("G" altered to "C," "A" altered to "T"). Generally, a single nucleotide change is sufficient to remove RM system recognition.
6. Once all proposed changes are introduced, repeat **step 2** to ensure that additional RM motifs have not been introduced during the process of RM silencing.
7. Synthesize the RM-silenced plasmid de novo using your preferred service provider. Using commercially available *Escherichia coli* strains, assemble and maintain the plasmid prior to transformation into *Fusobacterium* (*see* **Notes 14** and **15**).

3.4 Generating Electrocompetent Fusobacterium Cells

To be performed inside anaerobic chamber

1. Plate *Fusobacterium* strain on agar plate (*see* **Note 1**), and incubate anaerobically at 37 °C (*see* **Note 2**) for 24–72 h.

2. Using a sterile cotton swab, resuspend plate growth into 1 mL of pre-reduced, sterile tryptic soy broth (TSB) supplemented with 2.5% yeast extract and 0.4 mg/mL of L-cysteine hydrochloride (supplemented TSB).
3. Inoculate 100 mL of supplemented TSB with 25 μL of resuspended growth in a sterile 250-mL glass bottle.
4. Measure OD600nm.
5. Grow overnight for 23–24 h.
6. On day 2, pre-chill centrifuge to 4 °C, and place vials of 10% glycerol on ice to pre-chill. Glycerol will remain on ice throughout the entire protocol. When needed, place in fresh ice.
7. Measure OD600nm.
8. Aliquot 45 mL of growth into 50-mL sterile conical tubes for a total of two aliquots. Parafilm the top of each conical tube. Place tubes on ice and bring tubes out of the anaerobic chamber.
9. Centrifuge samples at 4 °C for 10 minutes at max speed. Place tubes back on ice.
10. Place tubes back into anaerobic chamber and pour off the supernatant.
11. Resuspend each pellet in 40 mL of pre-reduced, pre-chilled 10% glycerol. Parafilm the top of the conical tube. Placing tubes on ice, bring tubes out of the anaerobic chamber.
12. Centrifuge samples at 4 °C for 10 minutes at max speed. Place tubes back on ice.
13. Place tubes back into anaerobic chamber and pour off the supernatant.
14. Repeat **steps 11–13** using 20 mL of pre-reduced, pre-chilled 10% glycerol.
15. Repeat **steps 11–13** using 10 mL of pre-reduced, pre-chilled 10% glycerol.
16. Repeat **steps 11–13** using 2.25 mL of pre-reduced, pre-chilled 10% glycerol.
17. Pool both pellets (for a total of 4.5 mL) and transfer into a pre-chilled, sterile 5-mL Eppendorf tube. Place tubes on ice and bring tubes out of the anaerobic chamber.
18. Repeat **steps 12–13**.
19. Repeat **steps 11–13** using 1 mL of pre-reduced, pre-chilled 10% glycerol.
20. Centrifuge samples at 4 °C for 10 minutes at max speed. Place tubes back on ice.

21. Place tubes back into anaerobic chamber and pour off the supernatant.
22. Resuspend in the needed volume of 10% pre-reduced, pre--chilled glycerol to reach a desired final OD600 (*see* **Note 16**).
23. Aliquot 100 μL of competent cells into pre-chilled 1.5-mL Eppendorf tubes.
24. If cells will be immediately used for transformation, keep them on ice. Otherwise, freeze aliquots and store at −80 °C (*see* **Note 17**).

3.5 Electroporation of Plasmid Vector into Fusobacterium

To be performed inside anaerobic chamber and on ice (if possible), up until electroporation in ***step 6***

1. Add ~200 ng-1 μg of purified plasmid (*see* **Notes 18** and **19**).
2. Mix gently by pipetting, and transfer the entire volume into one pre-chilled electroporation cuvette. Close the lid tightly to avoid exposure to oxygen (*see* **Note 20**).
3. Outside the anaerobe chamber, electroporate at 2.0 kV (*see* **Note 19**).
4. Immediately add 1 mL of pre-warmed, pre-reduced, sterile tryptic soy broth (TSB) supplemented with 2.5% yeast extract and 0.4 mg/mL L-cysteine hydrochloride (supplemented TSB). Return to anaerobic chamber and incubate for approximately 2 h at 37 °C (*see* **Note 21**).
5. Plate a volume of the recovered bacteria onto an agar plate containing the appropriate antibiotic to facilitate plasmid selection (*see* **Notes 1**, **22**, and **23**).
6. Incubate the plate at 37 °C under anaerobic conditions. Colonies typically appear after 2–4 days.
7. Re-streak single colonies onto a fresh agar plate containing the appropriate antibiotic.
8. Verify plasmid presence by a method of choice (*see* **Note 24**).
9. Stock a positive colony in TSB and 20% glycerol solution (filter-sterilized). Store at −80 °C for future use.

4 Notes

1. We recommend using fastidious anaerobe agar plates supplemented with 10% defibrinated horse blood.
2. An anaerobic chamber with the capacity to incubate at 37 °C is preferred. In the absence of an anaerobic chamber, an AnaeroBox™ with its appropriate satchel will suffice.

3. We recommend pre-chilling your centrifuge during this step in order to be ready for **step 14** of procedure 3.1.
4. If you do not observe the formation of white spindles after inverting the tube 40 times, continue on to **step 23**. It is possible that the amount of precipitated DNA is small.
5. Alternatively, post-isopropanol precipitation samples can be centrifuged at 4 °C and at approximately 10,000 rcf for 2 minutes, isopropanol removed with a 1-mL pipette, and 1 mL of pre-chilled 70% ethanol can be added to dislodge and wash the pellet. However, we have observed higher concentrations and more optimal 260/280 and 260/230 values when spindles are removed and placed directly into pre-chilled 70% ethanol.
6. If unsure, start with a smaller volume. If after a 30-min incubation the DNA pellet does not dissolve, add more sterile water.
7. Due to the large size of gDNA, we recommend using the NEB 1 kb Extend DNA ladder on your gel.
8. REBASE summary files provide detailed insights on the functional roles of detected motifs. Note the motifs that are assigned to restriction-modification systems (which involve both restriction endonuclease and methyltransferase), orphan methyltransferases systems, and unassigned motifs (those that are actively methyl-modified but currently unassignable to a specific methyltransferase through REBASE).
9. Detailed annotations of genetic tools are important as changes will be introduced to optimize the plasmid for transformation, and ensuring these key elements are maintained or adjusted appropriately is crucial to maintain function. This can be achieved manually based on known features described within publications, through available NCBI annotations, or using tools such as PlasMapper 3.0 [11] or pLannotate [12], or VectorBuilder.
10. Understanding the distribution and position of RM motifs on plasmids or individual genetic parts aids in the design and optimization of RM-silent sequences.
11. Once all RM-associated motifs for a strain of interest have been identified and annotated, it is important to differentiate those that exist within coding regions compared to noncoding regions.
12. For example, in the target motif sequence M.FnuSB010II CTNNAG only nucleotide changes that disrupt the "C," "T," "A," or "G" will eliminate the RM-recognition motif. Modifications of the nucleotides at position 3 or 4 ("N") of this motif will not prevent restriction digestion from the corresponding restriction endonuclease.

13. The introduction of any single nucleotide modification in a DNA sequence can inadvertently introduce new RM recognition motif. It is important to query whether the altered nucleotide at each step introduces another recognition motif that will be targeted by an RM system within the strain of interest.
14. If all of the introduced changes are clustered to a discrete region of the plasmid, it may alternatively be possible to synthesize only the needed RM-silent fragment and assemble using the cloning strategy of your choice (Gibson, HiFi, Restriction Digest/Ligation, etc).
15. The inherent AT richness of DNA sequences associated with *Fusobacterium* can often lead to challenges in routine molecular procedures, including amplification, recombination, and plasmid assembly. This can extend to de novo DNA synthesis from commercial providers. It is critical to confirm the sequence of fragments and constructs at each step of the process as mutations can be introduced.
16. We recommend a resuspension volume for a final OD600nm = 60. This is calculated as (OD600nm of cells prior to cell preparation) × (preparation volume)/60. For example, if the OD600nm of the cells is 1.80 (**step** 7) and the preparation volume is 90 mL (**step 8**), then the resuspension volume = (1.80) × (90 mL)/60 = 2.7 mL.
17. Transformation with competent cells should be performed as quickly as possible and no later than 1 week after competent cell preparation.
18. The volume of plasmid added should be no larger than 1/10th of the volume of competent cells.
19. If larger amounts of plasmid are required or if the electroporation causes arcing, consider dialyzing the plasmid DNA (using, e.g., nylon membrane filter discs).
20. Avoid introduction of bubbles as this can lead to arcing during electroporation.
21. We recommend transferring the entire volume into a 5-mL round bottom tube and pulse spinning if possible. Otherwise, leaving the volume in the electroporation cuvette during incubation is acceptable.
22. The *catP* resistance cassette in combination with thiamphenicol is widely used for *Fusobacterium* plasmids. A working concentration of 5 μg/mL thiamphenicol is typically effective in nutrient rich agar plates like fastidious anaerobe agar (FAA) or brain heart infusion (BHI).
23. Depending on the nature of the plasmid sequence (replicative, transposon cassette, suicide), we recommend plating a range of volumes (e.g., 50 μL, 100 μL, and 250 μL) on different plates in cases where the transformation efficiency is unknown. In the

case of transposon mutagenesis, it is generally the case that all of the post-recovery volume is plated across different plates.

24. Most *Fusobacterium* replicons do not appear to support high copy numbers and standard plasmid extraction kits might yield lower-than-expected results. Direct colony sequencing or sequencing of PCR-amplicons often provide more reliable results.

Conflict of Interest Statement

C.D.J is an inventor on intellectual property related to Syngenic DNA technology licensed by Fred Hutchinson Cancer Center to Azitra Inc., which received compensation related to this license, and C.D.J serves as a consultant to Azitra Inc. This commercial entity had no role in this work. The remaining authors have no conflicts of interest to declare that are relevant to the content of this chapter.

References

1. Fan Z et al (2023) *Fusobacterium nucleatum* and its associated systemic diseases: epidemiologic studies and possible mechanisms. J Oral Microbiol 15(1):2145729. https://doi.org/10.1080/20002297.2022.2145729
2. Han YW (2015) *Fusobacterium nucleatum*: a commensal-turned pathogen. Curr Opin Microbiol 23:141–147. https://doi.org/10.1016/j.mib.2014.11.013
3. McIlvanna E, Linden GJ, Craig SG, Lundy FT, James JA (2021) *Fusobacterium nucleatum* and oral cancer: a critical review. BMC Cancer 21:1212. https://doi.org/10.1186/s12885-021-08903-4
4. Sun CH et al (2019) The role of *Fusobacterium nucleatum* in colorectal cancer: from carcinogenesis to clinical management. Chronic Dis Transl Med 5(3):178–187. https://doi.org/10.1016/j.cdtm.2019.09.001
5. Suzuki H (2012) Methylation-from DNA, RNA and histones to diseases and treatment. https://doi.org/10.5772/2932
6. Yasui K et al (2009) Improvement of bacterial transformation efficiency using plasmid artificial modification. Nucleic Acids Res 37(1):e3. https://doi.org/10.1093/nar/gkn884
7. Umana A et al (2022) Enhanced *Fusobacterium nucleatum* genetics using host DNA methyltransferases to bypass restriction-modification systems. J Bacteriol 204(12):e0027922. https://doi.org/10.1128/jb.00279-22
8. Johnston CD et al (2019) Systematic evasion of the restriction-modification barrier in bacteria. Proc Natl Acad Sci 116(23):11454–11459. https://doi.org/10.1073/pnas.1820256116
9. Eid J et al (2009) Real-time DNA sequencing from single polymerase molecules. Science 323(5910):133–138. https://doi.org/10.1126/science.1162986
10. Roberts RJ, Vincze T, Posfai J, Macelis D (2023) REBASE: a database for DNA restriction and modification: enzymes, genes and genomes. Nucleic Acids Res 51:D629–D630. https://doi.org/10.1093/nar/gkac975
11. Wishart DS et al (2023) PlasMapper 3.0-a web server for generating, editing, annotating and visualizing publication quality plasmid maps. Nucleic Acids Res 51:W459–W467. https://doi.org/10.1093/nar/gkad276
12. McGuffie MJ, Barrick JE (2021) pLannotate: engineered plasmid annotation. Nucleic Acids Res 49:W516–W522. https://doi.org/10.1093/nar/gkab374

Chapter 8

Tracking Surface Protein Assembly and Peptidoglycan Biosynthesis in *Fusobacterium nucleatum*

Timmie A. Britton, Bryan Sanchez, Asis Das, and Hung Ton-That

Abstract

Fusobacterium nucleatum is a Gram-negative, obligate anaerobe and oncogenic oral bacterium that employs a diverse array of surface-exposed adhesins to colonize extra-oral tissues. Among these, Fusobacterial adhesin A (FadA) is a well-characterized virulence factor known for its ability to target epithelial and endothelial tight junctions, disrupt barrier integrity, and activate pro-tumorigenic signaling pathways implicated in colorectal cancer. FadA also facilitates placental and fetal colonization, contributing to adverse pregnancy outcomes such as preterm birth and stillbirth. While its functional roles in bacterial colonization and pathogenicity are increasingly recognized, the spatial and temporal dynamics of FadA assembly on the bacterial surface—and its relationship to cell wall biogenesis—remain poorly understood. Here, we present a fluorescence-based labeling approach using fluorescent D-amino acids (FDAAs) to simultaneously track peptidoglycan biosynthesis and the localization of FadA in *F. nucleatum*. This methodology provides a versatile platform for investigating the dynamic interplay between virulence factor localization and cell envelope remodeling.

Key words Immunofluorescence, Fluorescently-tagged D-amino acids (FDAAs), HADA, Rf470DL, FadA, *Fusobacterium nucleatum*, Peptidoglycan, Gram-negative bacteria

1 Introduction

Fusobacterium nucleatum is a Gram-negative, anaerobic bacterium that has garnered significant attention for its role in periodontal disease and its emerging association with extraoral pathologies, including adverse pregnancy outcomes and colorectal, pancreatic, and breast cancers [1–4]. A key virulence factor of *F. nucleatum* is Fusobacterial adhesin A (FadA), a small protein that facilitates host cell attachment, invasion, and pro-tumorigenic signaling through interactions with epithelial and endothelial tight junction proteins [5–7]. Notably, FadA assembles into amyloid-like structures on the bacterial surface, which are essential for promoting colorectal cancer (CRC) progression in murine models [8]. Despite the clinical significance of FadA amyloidosis, the spatial and temporal dynamics

Hung Ton-That (ed.), *Fusobacterium nucleatum: Methods and Protocols*, Methods in Molecular Biology, vol. 3055,
https://doi.org/10.1007/978-1-0716-5388-3_8,

of FadA's surface localization and amyloid formation—and its relationship to cell wall biosynthesis—have remained poorly understood until recently [9].

The bacterial cell wall, composed primarily of peptidoglycan (PG), is a rigid, mesh-like structure that endows bacteria with their characteristic cell shapes and protects bacteria against osmotic stress. PG synthesis begins in the cytoplasm with the formation of lipid II, which is translocated across the cytoplasmic membrane to serve as a substrate for glycosyltransferases [10]. These enzymes polymerize glycan strands composed of repeating disaccharide units of N-acetylmuramic acid (NAM) and N-acetylglucosamine (NAG), each bearing peptide stems—pentapeptides in Gram-positive and tetrapeptides in Gram-negative bacteria—rich in D-amino acids [10]. Crosslinking of these stems via transpeptidation, catalyzed by penicillin-binding proteins (PBPs), confers structural integrity to the PG matrix [10].

Traditional methods for visualizing PG biosynthesis, such as radioactive labeling and fluorescent antibiotics (e.g., vancomycin-FL) or lectins (e.g., fluorescent wheat germ agglutinin), have provided valuable insights but suffer from limitations including poor membrane permeability as well as cytotoxicity [11]. These drawbacks have hindered our ability to study PG dynamics in live cells.

Fluorescently conjugated D-amino acids (FDAAs) offer a nontoxic, covalent labeling strategy to monitor PG synthesis in real time. Because D-amino acids are conserved in PG peptide stems, FDAAs serve as site-specific probes applicable across diverse bacterial species. The first FDAA, HADA (7-hydroxycoumarin-3-carboxylic acid conjugated to 3-amino-D-alanine), enabled visualization of PG incorporation at division septa and cell boundaries within a single generation time [12]. Its L-enantiomer, HALA, fails to label PG, confirming the transpeptidase specificity for D-amino acids [12]. HPLC and mass spectrometry analyses revealed minimal incorporation (0.2–2.8%) and exclusive labeling at terminal peptide positions, further supporting FDAA specificity and low toxicity [12].

Since their development in 2012, over a dozen FDAA probes have been synthesized, spanning emission spectra from blue (HADA) to far-red (RADA, Rf470DL, RMR-Tre). FDAA labeling reflects stem peptide remodeling and has been used to identify active PG synthesis zones. Short-pulse labeling highlights newly synthesized PG and has revealed growth patterns in *Streptomyces venezuelae* and *Lactobacillus lactis* [11]. Long-pulse labeling, akin to pulse-chase experiments, has been used to assess PG turnover in *Streptococcus pneumoniae* [13]. Sequential labeling with multiple FDAAs and time-lapse microscopy has enabled real-time tracking of PG dynamics, demonstrating roles for cytoskeletal proteins like FtsZ in *Staphylococcus aureus* and *Bacillus subtilis* [14, 15], and

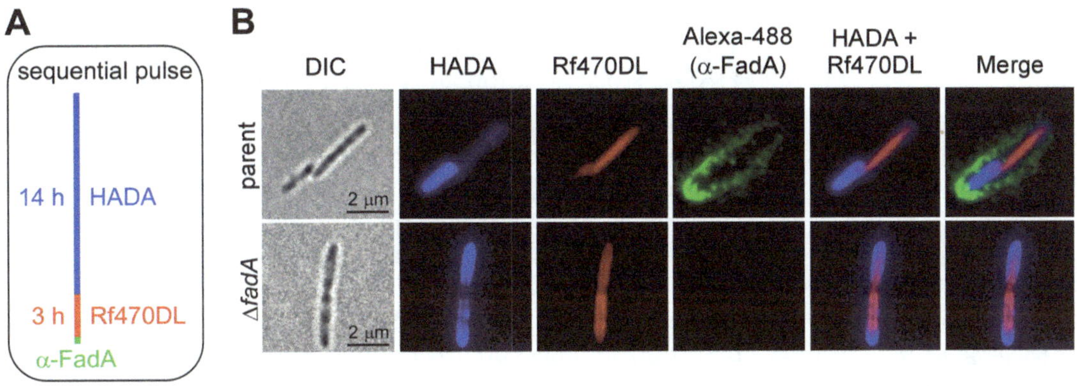

Fig. 1 Sequential labeling and immunofluorescence microscopy of FadA. (**A-B**) Overnight cultures of parent and Δ*fadA* mutant strains were diluted (OD_{600} 0.05) and sequentially labeled with fluorescently conjugated D-amino acid dyes, HADA (blue) and Rf470DL (red), for 14 h and 3 h, respectively. Afterward, cells were harvested and washed in PBS before labeling with α-FadA and Alexa488-conjugated secondary antibodies (green). Samples were analyzed by fluorescence microscopy. All results were obtained from three independent experiments performed in triplicate

revealing septal enrichment of the signal peptidase SpsB in *S. aureus* [16].

Here, we present a sequential FDAA labeling protocol combined with immunofluorescence microscopy (IFM) to investigate the spatiotemporal dynamics of FadA in *F. nucleatum*, based on our previous publication. Our findings suggest that FadA localizes predominantly to the old cell pole, indicating that its distribution may be influenced by the bacterium's growth state (*see* Fig. 1) [9]. This protocol detailed here provides a powerful tool for studying PG biosynthesis and cell wall-associated protein localization in *F. nucleatum*, a key player in both oral and systemic disease.

2 Materials

All solutions should be prepared with molecular-grade sterile water.

2.1 Coverslip Preparation

1. No. 1 coverslip (12 mm diameter).
2. 0.01% poly-L-lysine solution: prepare from lyophilized poly-L-lysine hydrobromide powder, and dissolve in water.

2.2 Preparation of F. nucleatum Cells for Sequential Labeling

1. *Fusobacterium nucleatum* strains: ATCC 23726, CW1 [17], Δ*rnfC* [18], Δ*rnfC*/pRnfC [18], and Δ*fadA* [9].
2. TSPC: Tryptic Soy Broth (TSB) supplemented with 1% Bacto peptone and 0.25% freshly prepared L-cysteine.
3. 1.5-mL Eppendorf tubes.
4. Phosphate-buffered saline (PBS), pH 7.4.

5. 3-[[(7-Hydroxy-2-oxo-2*H*-1-benzopyran-3-yl)carbonyl] amino]-D-alanine hydrochloride (HADA): dilute 50 mM of HADA stock solution in TSPC to a final concentration of 0.5 mM.
6. N^6-[3-[6-[5-[2-Carboxy-2-cyanoethenyl]-2-thienyl]-3,4-dihydro-1 (2*H*)-quinolinyl]-1-oxopropyl]-D-lysine hydrochloride (Rf470DL): dilute 50 mM of Rf470DL stock solution in TSPC to a final concentration of 0.5 mM.

2.3 Permeabilization of Bacterial Cells

1. 4% formaldehyde solution: dissolve 4 grams of paraformaldehyde powder into 100 mL 1X PBS. Filter the solution through a 0.22-μm sterile filter into a sterile container.
2. 70% ethanol.
3. TEG buffer: 25 mM of Tris-HCl, pH 8.0, 10 mM of EDTA, and 50 mM of glucose.
4. 50 μg/mL of lysozyme solution in TEG buffer.
5. End-to-end rotator.

2.4 Immunolabeling of Bacterial Cells

1. 3% bovine serum albumin (BSA) solution: dissolve 3 grams of BSA powder into 100 mL 1X PBS (sterile filtration before use).
2. 4% paraformaldehyde in PBS: dissolve 4 grams of paraformaldehyde powder into 100 mL 1X PBS (sterile filtration before use).
3. Primary antibody: rabbit anti-FadA (α-FadA).
4. Secondary antibody: goat anti-rabbit IgG AlexaFluor 488 conjugate.

2.5 Mounting and Fluorescence Microscopy

1. Microscope slides (25 mm × 75 mm × 1 mm).
2. Nail polish.
3. Dumont angled tweezers.
4. A fluorescence microscope with appropriate fluorescence filters.

3 Methods

3.1 Coverslip Preparation

1. Apply 150 μL of poly-L-lysine solution to one side of the coverslips and let settle for at least 5 min (*see* **Note 1**).
2. Aspirate poly-L-lysine from coverslips and rinse coated surface thoroughly with sterile molecular grade water.
3. Allow coverslips to air-dry for at least 2 h (*see* **Note 2**).

3.2 Preparation of F. nucleatum Cells for Sequential Labeling

1. Aseptically transfer a single colony of individual *F. nucleatum* strains to 5 mL of TSPC and grow cultures at 37 °C overnight in an anaerobic chamber (10% CO_2, 10% H_2, and 80% N_2).
2. Adjust the density of fusobacterial cell cultures to OD_{600} of 0.05 in 1.5-mL Eppendorf tubes containing TSPC with a final HADA concentration of 0.05 mM. Grow cultures at 37 °C for 14 h in the anaerobic chamber.
3. Harvest cells by centrifugation at 10,000 × g for 2 min, followed by aspiration of bacterial supernatants. Wash cell pellets once in PBS.
4. Adjust the density of fusobacterial cell cultures to OD_{600} of 0.05 in 1.5-mL Eppendorf tubes containing TSPC with a final Rf470DL concentration of 0.05 mM. Grow cultures at 37 °C for 3 h in the anaerobic chamber.
5. Harvest cells by centrifugation at 10,000 × g for 2 min, followed by aspiration of bacterial supernatants. Wash cell pellets twice in PBS.
6. For labeling of integral membrane proteins, proceed to Subheading 3.3 for permeabilization. For labeling of peripheral and/or surface-exposed proteins, proceed to Subheading 3.4.

3.3 Permeabilization of Bacterial Cells

1. Resuspend pellets obtained from **step 5** of Subheading 3.2 in 1 mL of 4% formaldehyde solution and incubate for 30 min at room temperature (*see* **Note 3**).
2. Pellet cells by centrifugation at 6000 × g for 2 min and wash three times in PBS.
3. Resuspend cell pellets in 1 mL of 70% ethanol, and incubate at room temperature for 1 h on an end-to-end rotator.
4. Harvest cells by centrifugation at 6000 × g for 2 min and discard supernatant.
5. Resuspend cell pellets in 0.5 mL of TEG buffer, add 0.5 mL of 50 μg/mL lysozyme solution in TEG, and incubate at room temperature for 30 min.
6. Transfer 150 μL of cell suspension to poly-L-lysine-coated coverslips from Subheading 3.1 and incubate at room temperature for 5 min.
7. Aspirate the remaining solution from the coverslips carefully, and gently wash coverslip with 150 μL of PBS to remove unbound cells.
8. Proceed to Subheading 3.4.

3.4 Immunolabeling of Bacterial Cells

1. Transfer 150 μL of cell suspensions to labeled, poly-L-lysine-coated coverslips (*see* Subheading 3.1). Incubate at room temperature for 15 min.

2. Carefully aspirate the remaining cell suspension, and wash coverslips with 150 μL of PBS.
3. To fix cells, apply 150 μL of 4% formaldehyde solution onto the coverslip and incubate for 20 min at room temperature (*see* **Note 3**).
4. Block cells on coverslip with 150 μL of 3% BSA solution at room temperature for at least 1 h. Afterward, aspirate the BSA solution.
5. Dilute primary antibody in 3% BSA (α-FadA 1:3000), and add 150 μL onto coverslips. Incubate at room temperature for 1 h.
6. Wash coverslips three times with PBS (*see* **Note 4**).
7. Dilute secondary antibody in 3% BSA (α-Rabbit, 1:10,000), and add 150 μL onto coverslips. Cover and incubate at room temperature for 1 h. Keep coverslips in the dark for all remaining steps to prevent photobleaching (*see* **Note 5**).
8. Wash coverslips with 150 μL of PBS.

3.5 Mounting and Fluorescence Microscopy

1. Aspirate any remaining PBS from coverslips.
2. Using Dumont angled tweezers, invert coverslip and mount face down onto a microscopy slide.
3. Carefully apply nail polish along the perimeter of the coverslip to seal. Let dry at room temperature for 5 min (*see* **Note 6**).
4. Image via a fluorescence microscope, using the appropriate channel for the indicated excitation and emission spectra of fluorophore of interest (*see* **Note 7**).

4 Notes

1. The incubation of coverslips with poly-L-lysine overnight enhances bacterial cell attachment.
2. Dry coverslip under flame for 2 h or cover and air-dry overnight.
3. Store 4% formaldehyde solution at 4 °C.
4. When washing the coverslips with PBS, pipette up and down gently to preserve surface tension. Avoid touching coverslip. Incubate for at least 5 min prior to aspirating.
5. Researchers can either move coverslips into a dark room after the addition of Alexa488-conjugated secondary antibody, or simply cover the coverslip with a small Eppendorf box.
6. Stained microscope slides can be stored at room temperature. Storage at −20 °C may create moisture upon thawing, which may damage the integrity of the nail polish seal. Protect from light.

7. For AlexaFluor 488, use the green fluorescence channel (max excitation and emission at 490 nm and 525 nm, respectively). For HADA, use the blue fluorescence channel (max excitation and emission at 405 nm and 450 nm, respectively). For Rf470DL, use the red fluorescence channel (max excitation and emission at 470 nm and 640 nm, respectively).

Acknowledgments

We would like to thank our laboratory members for their critical review of this manuscript. Work reported in this publication was supported by the National Institute of Dental and Craniofacial Research (NIDCR) of the National Institutes of Health (NIH) under the award number DE026758 (to H.T.T.). The content presented in this publication does not reflect the viewpoints of the National Institutes of Health and is solely the responsibility of the authors.

Conflict of Interest Statement The authors have no conflicts of interest to declare that are relevant to the content of this chapter.

References

1. Han YW, Shen T, Chung P, Buhimschi IA, Buhimschi CS (2009) Uncultivated bacteria as etiologic agents of intra-amniotic inflammation leading to preterm birth. J Clin Microbiol 47(1):38–47. https://doi.org/10.1128/JCM.01206-08
2. Payne MS, Bayatibojakhi S (2014) Exploring preterm birth as a polymicrobial disease: an overview of the uterine microbiome. Front Immunol 5:595. https://doi.org/10.3389/fimmu.2014.00595
3. Parhi L, Alon-Maimon T, Sol A, Nejman D, Shhadeh A, Fainsod-Levi T, Yajuk O, Isaacson B, Abed J, Maalouf N, Nissan A, Sandbank J, Yehuda-Shnaidman E, Ponath F, Vogel J, Mandelboim O, Granot Z, Straussman R, Bachrach G (2020) Breast cancer colonization by *Fusobacterium nucleatum* accelerates tumor growth and metastatic progression. Nat Commun 11(1):3259. https://doi.org/10.1038/s41467-020-16967-2
4. Udayasuryan B, Ahmad RN, Nguyen TTD, Umana A, Monet Roberts L, Sobol P, Jones SD, Munson JM, Slade DJ, Verbridge SS (2022) *Fusobacterium nucleatum* induces proliferation and migration in pancreatic cancer cells through host autocrine and paracrine signaling. Sci Signal 15(756):eabn4948. https://doi.org/10.1126/scisignal.abn4948
5. Han YW, Ikegami A, Rajanna C, Kawsar HI, Zhou Y, Li M, Sojar HT, Genco RJ, Kuramitsu HK, Deng CX (2005) Identification and characterization of a novel adhesin unique to oral fusobacteria. J Bacteriol 187(15):5330–5340. https://doi.org/10.1128/JB.187.15.5330-5340.2005
6. Rubinstein MR, Wang X, Liu W, Hao Y, Cai G, Han YW (2013) *Fusobacterium nucleatum* promotes colorectal carcinogenesis by modulating E-cadherin/beta-catenin signaling via its FadA adhesin. Cell Host Microbe 14(2):195–206. https://doi.org/10.1016/j.chom.2013.07.012
7. Fardini Y, Wang X, Temoin S, Nithianantham S, Lee D, Shoham M, Han YW (2011) *Fusobacterium nucleatum* adhesin FadA binds vascular endothelial cadherin and alters endothelial integrity. Mol Microbiol 82(6):1468–1480. https://doi.org/10.1111/j.1365-2958.2011.07905.x
8. Meng Q, Gao Q, Mehrazarin S, Tangwanichgapong K, Wang Y, Huang Y, Pan Y, Robinson S, Liu Z, Zangiabadi A, Lux R, Papapanou PN, Guo XE, Wang H,

Berchowitz LE, Han YW (2021) *Fusobacterium nucleatum* secretes amyloid-like FadA to enhance pathogenicity. EMBO Rep 22(7): e52891. https://doi.org/10.15252/embr.202152891

9. Britton TA, Lee JH, Chang C, Bhat AH, Chen YW, Mohammed Ali R, Wu C, Das A, Ton-That H (2025) Inactivation of the *Fusobacterium nucleatum* Rnf complex reduces FadA-mediated amyloid formation and tumor development. MBio 16(6):e0103225. https://doi.org/10.1128/mbio.01032-25
10. Garde S, Chodisetti PK, Reddy M (2021) Peptidoglycan: structure, synthesis, and regulation. EcoSal Plus 9(2). https://doi.org/10.1128/ecosalplus.ESP-0010-2020
11. Hsu YP, Booher G, Egan A, Vollmer W, VanNieuwenhze MS (2019) d-amino acid derivatives as in situ probes for visualizing bacterial peptidoglycan biosynthesis. Acc Chem Res 52(9):2713–2722. https://doi.org/10.1021/acs.accounts.9b00311
12. Kuru E, Hughes HV, Brown PJ, Hall E, Tekkam S, Cava F, de Pedro MA, Brun YV, VanNieuwenhze MS (2012) In Situ probing of newly synthesized peptidoglycan in live bacteria with fluorescent D-amino acids. Angew Chem Int Ed Engl 51(50):12519–12523. https://doi.org/10.1002/anie.201206749
13. Boersma MJ, Kuru E, Rittichier JT, MS VN, Brun YV, Winkler ME (2015) Minimal peptidoglycan (PG) turnover in wild-type and PG hydrolase and cell division mutants of Streptococcus pneumoniae D39 growing planktonically and in host-relevant biofilms. J Bacteriol 197(21):3472–3485. https://doi.org/JB.00541-15
14. Pereira AR, Hsin J, Krol E, Tavares AC, Flores P, Hoiczyk E, Ng N, Dajkovic A, Brun YV, MS VN, Roemer T, Carballido-Lopez R, Scheffers DJ, Huang KC, Pinho MG (2016) FtsZ-dependent elongation of a coccoid bacterium. mBio 7(5) https://doi.org/mBio.00908-16
15. Bisson-Filho AW, Hsu YP, Squyres GR, Kuru E, Wu F, Jukes C, Sun Y, Dekker C, Holden S, VanNieuwenhze MS, Brun YV, Garner EC (2017) Treadmilling by FtsZ filaments drives peptidoglycan synthesis and bacterial cell division. Science 355(6326):739–743. https://doi.org/355/6326/739
16. Zhang R, Jia Y, Scaffidi SJ, Madsen JJ, Yu W (2024) Signal peptidase SpsB coordinates staphylococcal cell cycle, surface protein septal trafficking, and LTA synthesis. mBio 16(3): e0267324. https://doi.org/mbio02673-24
17. Wu C, Al Mamun AAM, Luong TT, Hu B, Gu J, Lee JH, D'Amore M, Das A, Ton-That H (2018) Forward genetic dissection of biofilm development by *Fusobacterium nucleatum*: novel functions of cell division proteins FtsX and EnvC. mBio 9(2) https://doi.org/mBio.00360-18
18. Britton TA, Wu C, Chen YW, Franklin D, Chen Y, Camacho MI, Luong TT, Das A, Ton-That H (2024) The respiratory enzyme complex Rnf is vital for metabolic adaptation and virulence in *Fusobacterium nucleatum*. mBio 15(1):e0175123. https://doi.org/01751-23

Chapter 9

Isolation of *Fusobacterium nucleatum*-Induced Host Exosomal sRNAs

Ajay Kumar and Xuesong He

Abstract

In eukaryotic cells, small RNAs (sRNAs) are a class of noncoding RNAs that are less than 200 nucleotides in length and mainly classified per their biogenesis as microRNA, small interfering RNA, PIWI-interacting RNA (piRNA) or tRNA-derived small RNA (tsRNA). Increasing lines of evidence indicate that these host-derived sRNAs may contribute to the maintenance of host-microbial homeostasis through cross-kingdom gene modulation. Recent studies have shown that when challenged with *Fusobacterium nucleatum*, a Gram-negative, anaerobic oral pathobiont, host epithelial cells release specific exosome-borne sRNAs, particularly tsRNAs that specifically target *F. nucleatum* and modulate its growth. Here, we provide an experimental procedure to isolate exosomal host-derived sRNAs when challenged with *F. nucleatum*. In principle, this protocol can be used for any in vitro system to study bacterial host interactions.

Key words *Fusobacterium nucleatum*, Exosome, Gram-negative bacteria, Small RNAs

1 Introduction

Multiple innate and adaptive immune responses involving antimicrobial peptides, complement, and immunoglobulins have evolved to maintain the delicate balance between the host and associated microbiomes [1–3]. In addition to these well-established systems, recent studies have begun to shed light on the roles of host-derived small RNAs (sRNAs) that contribute to the maintenance of host-microbial homeostasis through cross-kingdom gene modulation [4, 5]. Among various host-derived sRNA types, tsRNAs are a distinct class of small noncoding RNAs derived from the cleavage of mature tRNAs [6]. Though initially thought to be degradation products, recent research has revealed their involvement in various cellular processes, including gene regulation, immune modulation, and antimicrobial defense [6]. The ability of tsRNAs to target bacterial species and influence their growth suggests they serve as an intrinsic defense mechanism against microbial infections. These

Hung Ton-That (ed.), *Fusobacterium nucleatum: Methods and Protocols*, Methods in Molecular Biology, vol. 3055,
https://doi.org/10.1007/978-1-0716-5388-3_9,

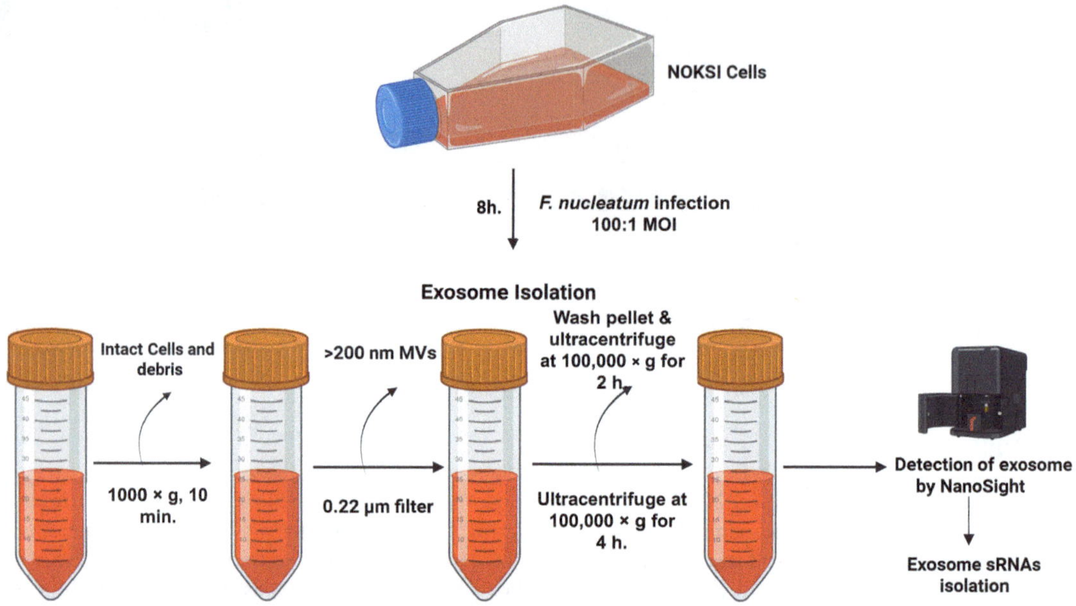

Fig. 1 Overview of exosome isolation. After NOSKI cells reach full growth and are infected with *Fusobacterium nucleatum*, the culture media is collected. Exosomes are then isolated using centrifugation and ultracentrifugation. Finally, exosomes are analyzed by NanoSight and small RNA isolated

host derived tsRNAs can be packaged together with other sRNA types in exosomes [7].

Fusobacterium nucleatum is a Gram-negative, anaerobic pathobiont primarily found in the human oral microbiome. It has been implicated in various intraoral and extraoral diseases, including periodontitis [8], colorectal cancer, preterm delivery, and other systemic inflammatory diseases [9, 10]. Using a normal oral keratinocyte-spontaneously immortalized (NOKSI)-*F. nucleatum* in vitro host-microbial interacting system, He et al. demonstrated that when challenged with *F. nucleatum*, NOKSI cells released specific exosome-borne tsRNAs [7]. These tsRNAs display selective antimicrobial activity against *F. nucleatum*. Due to its pathogenic potential and involvement in various diseases, targeting *F. nucleatum* has become a key area of research. Thus, identifying host-derived tsRNAs that specifically target this pathogen could pave the way for innovative therapeutic strategies to control its growth and its role in disease progression.

Using NOKSI- *F. nucleatum* in vitro host-microbial interacting system as an experimental model, we provide a detailed procedure to isolate host-derived exosomal sRNAs when challenged with bacteria (*see* Fig. 1). The isolated sRNAs can be further characterized/categorized by sequencing and bioinformatic analysis and used for (1) investigating bacteria-induced host exosomal sRNA and (2) identifying host-derived tsRNAs (or other sRNA types) that may play a role in targeted modulation of bacterial physiology.

2 Materials

This comprehensive list of materials and reagents is essential for conducting experiments aimed at isolating *tsRNAs*, providing the necessary tools to explore their potential roles in targeting *F. nucleatum* effectively.

2.1 Bacterial Growth Preparation

1. *F. nucleatum* strain ATCC 23726 frozen stock.
2. Autoclave.
3. Brain heart infusion (BHI) broth media composition for 1 liter: 7.7 g of calf brains infusion, 9.8 g of beef heart infusion, 10 g of proteose peptone, 2 g of dextrose, 5 g of sodium chloride, and 2.5 g of disodium phosphate.
4. Yeast extract.
5. 5 mg/mL stock of hemin.
6. 1 mg/mL stock of menadione.
7. Anaerobic chamber (5% hydrogen, 10% carbon dioxide, and 85% nitrogen).

2.2 Cell Culture Preparation

1. Normal oral keratinocyte-spontaneously immortalized (NOKSI) cell line.
2. 37 °C water bath.
3. KSM (keratinocyte serum free medium) culture medium.
4. Cell culture CO_2 incubator.
5. Centrifuge.
6. Accutase solution (1× Accutase enzymes in PBS, 0.5 mM of EDTA).
7. T25 and T175 cell culture flasks.

2.3 Exosome Isolation Preparation

1. Polycarbonate ultracentrifuge tubes (70 mL).
2. Beckman coulter ultracentrifuge.
3. Pre-cooled type 45 Ti fixed-angle rotor at 4 °C overnight.
4. Nano-sight.
5. 0.22-μm filtration units.
6. Invitrogen total exosome RNA and protein isolation kits.

3 Methods

3.1 *F. nucleatum* Culture

1. Inoculate *F. nucleatum* from frozen stock in a 15-mL conical tube containing 1 mL of autoclaved brain heart infusion (BHI) broth media, supplemented with yeast extract (5 mg/mL), hemin (5 μg/mL), and menadione (0.5 μg/mL).
2. Incubate *F. nucleatum* liquid culture until it reaches the exponential growth phase in a 37 °C anaerobic chamber with 85% N_2, 10% CO_2, and 5% H_2.

3.2 NOKSI Cell Line Culture and Infection with *F. nucleatum*

1. Take out the stored NOKSI cell line vial from liquid nitrogen, and rapidly thaw by gentle agitation in a pre-heated 37 °C water bath. To reduce the possibility of contamination, keep the O-ring and cap out of the water.
2. After thawing, immediately transfer the vial culture to a 15-mL centrifuge tube containing 9.0 mL of pre-warmed complete KSM culture medium (*see* **Note 1**).
3. Spin at approximately 125 × g for 5 min at room temperature (RT).
4. Discard the supernatant and carefully resuspend the pellet in 5 mL of fresh KSM medium.
5. Transfer the resuspended 5-mL medium into a T25 flask, and incubate in a 37 °C cell culture CO_2 incubator until reaching 90% confluency.
6. After confluency, discard the culture medium.
7. Rinse the confluent cell layer twice for 10–15 s with 10–15 mL of sterile PBS to remove all dead cells.
8. Add 2.0 to 3.0 mL of Accutase solution to the confluent cells in the T25 flask, and incubate for 10–15 min in a CO_2 37 °C incubator until the cell layer is dispersed (observe under an inverted microscope) (*see* **Note 2**).
9. After detachment of cells, directly add 5 mL of new complete KSM medium, and aspirate cells by gently pipetting.
10. Add appropriate aliquots of the cell suspension to new culture vessels. Cultures can be established between 1.0×10^4 and 3.0×10^4 viable cells/cm^2 and incubated at 37 °C in a CO_2 incubator (*see* **Note 3**).
11. To scale up the culture, first, determine the number of cells using an automated cell counter or hemocytometer.
12. Add appropriate aliquots (4.9×10^6 cells/flask) of the viable cell suspension to a T175 culture flask (*see* **Note 4**).
13. Incubate cultures at CO_2 37 °C incubator, and take 3–4 days to reach above 90% confluency.

14. Aspirate the KSM medium from an 80–90% confluent T175 flask of cells and rinse two times with 25–30 mL of sterile RT PBS. The cells are now ready for *F. nucleatum* infection (*see* **Note 5**).
15. Replace medium with 50 mL of fresh complete KSM medium into each T175 flask.
16. Infect with exponential phase *F. nucleatum* at 100:1 MOI, and incubate in a cell culture incubator (at 37 °C, 5% CO_2) for a minimum of 8 h.

3.3 Exosomes Isolation

1. Prior to isolation, use sterile deionized and distilled water to wash the polycarbonate ultracentrifuge tubes and their covers (*see* **Note 6**).
2. Pour 70% ethanol into the ultracentrifuge tubes, and keep for 30 min.
3. Flip tubes over every 5–10-min interval.
4. Transfer tubes to a biosafety cabinet and remove ethanol. Allow the tubes and their lids dry in the air.
5. After 8-h incubations, carefully collect the cell supernatant from the T175 flasks (prepared in Subheading 3.2, **step 16**) using a serological pipette and place in 50-mL conical tubes on ice.
6. Centrifuge collected supernatants at low speed (1000 × g for 10 min at 4 °C) to separate intact cells and cell debris from the medium (*see* **Note** 7).
7. Collect the supernatant in a new 50-mL conical tube without disturbing the pellet (could leave 1 mL to avoid pellet disturbance). The collected supernatant now contains soluble growth factors, cytokines, macro vesicles, and exosomes [11].
8. Pass filtrate through a 0.22-μm filter to remove macrovesicles from the collected supernatant (*see* **Note 8**).
9. Rinse the filter with 5 mL of PBS, and apply the vacuum to remove any residual medium and exosomes from the filter.
10. When all the PBS has passed through the filter, immediately disconnect the vacuum tubing from the filter unit, and unscrew the filter from the filter.
11. Place the filtrate on ice. Ultracentrifugation will now be used to isolate the exosomes from the filtrate.
12. Pour the filtrate into polycarbonate ultracentrifuge tubes (prepared in **steps 1–4** of this section), and normalize tube weight using sterile PBS.
13. Place the ultracentrifuge tubes into a pre-cooled Type 45 Ti fixed-angle rotor (*see* **Note 9**).
14. Spin samples at 100,000 × g, at 4 °C for 4 h.

15. After centrifugation, carefully remove the tubes from the rotor, and place them on ice.
16. Remove the lids from the ultracentrifuge tubes. Without disturbing the pellet, carefully aspirate the supernatant into a pre-chilled 50-mL conical tube using a 25-mL serological pipette (*see* **Note 10**).
17. Keep tube on ice and circle the pellet with a marker.
18. Wash the exosome pellet by resuspending in 30 mL of ice-cold sterile PBS.
19. Weigh the ultracentrifuge tubes again, and normalize their weight using sterile PBS.
20. Centrifuge the samples at 100,000 × g, 4 °C for 2 h.
21. Remove the PBS and resuspend the pellet in 250 μL of ice-cold PBS (extensive pipetting).
22. Store at −80 °C for further use.
23. Isolate exosome RNA from samples using an Invitrogen total exosome RNA and protein isolation kit.
24. Confirm exosome sRNA isolation by small RNA-Seq analysis.

3.4 Characterization of Exosomes by NanoSight

Exosome characterization by NanoSight (Nanoparticle Tracking Analysis, NTA) involves exosome sample dilution in filtered sterile PBS to achieve an optimal particle concentration.

1. Aspirate around 500 μL of exosome sample, and load into the NanoSight pro (Malvern Panalytical) chamber using a syringe pump for uniform flow.
2. Adjust the laser and camera settings to optimize particle visualization.
3. Record multiple 30- to 60-s videos to capture the Brownian motion of particles.
4. The software tracks individual particle movement to calculate size distribution based on the Stokes-Einstein equation. Determine particle size (mode and mean diameter) and concentration.
5. Confirm the expected size range (~30–200 nm for exosomes).
6. Compare concentration and size distribution across different samples or conditions.

4 Notes

1. Keratinocyte SFM is a complete serum-free medium supplemented with human recombinant epidermal growth factor (rEGF) and bovine pituitary extract (BPE) at the time of use.

This specialized medium supports optimal cell growth and proliferation, providing essential nutrients.

2. When adding Accutase to the cells, do not agitate them by hitting or shaking the flask while waiting for the cells to detach. Difficult-to-detach cells may be placed at 37 °C to facilitate dispersal. After detaching, it is not necessary to eliminate the accutase solution; one may directly introduce the culture media and continue with subsequent procedures. When adding the culture media, ensure that it is done gently to prevent any potential damage to the cells and maintain their viability for further experimentation.
3. Ensuring optimal growth conditions and preventing overcrowding can lead to nutrient depletion and hinder cell proliferation. Sub-cultivation ratio: It is advisable to employ a sub-cultivation ratio within the parameters of 1:3 to 1:8. Regular monitoring of the cultures is essential to assess their health.
4. To get a good number of exosomes to isolate transfer RNA fragments, utilizing a minimum of two T175 flasks containing cells at a confluency above 80% is advisable.
5. All rinsing steps should be performed gently so as not to cause cell detachment. Maintaining a sterile environment throughout the process is essential, minimizing the risk of contamination that could compromise the integrity of the cultures and skew experimental results.
6. Nuclease-free water is used to prepare reagents, wash the steps, and sterilize the ultracentrifuge tubes/lids.
7. Every centrifugation step should be conducted at 4 °C. The reagents and harvested samples must be kept on ice.
8. To achieve effective filtration, it is essential to maintain a low vacuum pressure, allowing the filtration to flow through the membrane in drops; an overly strong vacuum may result in the passage of MVs through the filter. Every procedure must be executed within a sterile tissue culture hood.
9. Before inserting the ultracentrifuge tubes into the rotor, it is beneficial to mark the sides of the tubes that will be oriented towards the outer edge of the rotor. This will assist in identifying the location of the exosome pellets post-centrifugation.
10. The exosome pellet will settle at the base of the tube, beneath the marked line. It can be challenging to detect exosome pellets. Illuminating from below can aid in their visibility. At this stage, a slightly off-white to clear exosome pellet will be noticeable on the lower wall of the side marked on the ultracentrifuge tube.

Acknowledgments

This work was supported by NIH National Institute of Dental and Craniofacial Research (NIDCR) awards, DE030943 (XH), DE023810 (XH), and DE029479 (XH). We are grateful to our lab manager, Cen Lujia, and Dr. Mary Ellen Davey and Dr. Hyun Young Kim for providing technical support for using the NanoSight.

Conflict of Interest The authors have no conflicts of interest to declare that are relevant to the content of this chapter.

References

1. Chehoud C et al (2013) Complement modulates the cutaneous microbiome and inflammatory milieu. Proc Natl Acad Sci 110(37): 15061–15066. https://doi.org/10.1073/pnas.1307855110
2. Diamond G, Beckloff N, Weinberg A, Kisich KO (2009) The roles of antimicrobial peptides in innate host defense. Curr Pharm Des 15(21):2377–2392. https://doi.org/10.2174/138161209788682325
3. Sawa T, Kinoshita M, Inoue K, Ohara J, Moriyama K (2019) Immunoglobulin for treating bacterial infections: one more mechanism of action. Antibodies (Basel) 8(4):52. https://doi.org/10.3390/antib8040052
4. Cai Q, He B, Kogel KH, Jin H (2018) Cross-kingdom RNA trafficking and environmental RNAi-nature's blueprint for modern crop protection strategies. Curr Opin Microbiol 46:58–64. https://doi.org/10.1016/j.mib.2018.02.003
5. Zeng J et al (2019) Cross-kingdom small RNAs among animals, plants and microbes. Cells 8(4):371. https://doi.org/10.3390/cells8040371
6. Li S, Xu Z, Sheng J (2018) tRNA-derived small RNA: a novel regulatory small non-coding RNA. Genes (Basel) 9(5):246. https://doi.org/10.3390/genes9050246
7. He X et al (2018) Human tRNA-derived small RNAs modulate host-oral microbial interactions. J Dent Res 97(11):1236–1243. https://doi.org/10.1177/0022034518770605
8. Bolstad AI, Jensen HB, Bakken V (1996) Taxonomy, biology, and periodontal aspects of Fusobacterium nucleatum. Clin Microbiol Rev 9(1):55–71. https://doi.org/10.1128/CMR.9.1.55
9. Shang FM, Liu HL (2018) Fusobacterium nucleatum and colorectal cancer: a review. World J Gastrointest Oncol 10(3):71–81. https://doi.org/10.4251/wjgo.v10.i3.71
10. Shigefuku R et al (2017) Fusobacterium nucleatum detected simultaneously in a pyogenic liver abscess and advanced sigmoid colon cancer. Anaerobe 48:144–146. https://doi.org/10.1016/j.anaerobe.2017.08.010
11. Wang F, Cerione RA, Antonyak MA (2021) Isolation and characterization of extracellular vesicles produced by cell lines. STAR Protoc 2(1):100295. https://doi.org/10.1016/j.xpro.2021.100295

Chapter 10

Isolation of Primary RNA for Differential RNA Sequencing in *Fusobacterium nucleatum*

Falk Ponath and Jörg Vogel

Abstract

Understanding gene expression in *Fusobacterium nucleatum* and related species is essential for uncovering the molecular mechanisms that underlie their adaption to various environments. However, the isolation of high-quality RNA from these Gram-negative anaerobes is challenging because of their cell wall structure and the limitations of conventional column-based RNA purification kits, particularly their transcript size bias. Here, we present a robust, phenol-based RNA extraction protocol optimized for *F. nucleatum* that consistently yields high-integrity, DNA-free RNA suitable for transcriptomic analyses, including RNA sequencing (RNA-seq) and reverse transcription quantitative PCR (RT-qPCR). This method enables effective cell lysis, nucleic acid purification, and efficient removal of genomic DNA while preserving all transcript classes, including small regulatory RNAs (sRNAs). This protocol enabled us to successfully isolate RNA from different *F. nucleatum* strains and the related species *Fusobacterium periodonticum* and lead to the mapping of transcriptional starts sites and identification of sRNAs. Thus, this protocol provides a basis for future studies investigating the gene expression dynamics and regulatory networks of *F. nucleatum* under diverse environmental conditions.

Key words RNA, *Fusobacterium*, RNA sequencing, Small regulatory RNAs, RNA isolation

1 Introduction

Investigating global gene expression in medically relevant bacterial species can provide valuable insights into their physiology and capability to colonize and survive in or on the human host. A critical step in transcriptome-focused gene expression analysis such as RNA-sequencing (RNA-seq) or reverse transcription quantitative PCR (RT-qPCR) is to obtain high-quality RNA. While commercial RNA isolation kits offer a convenient solution, they are typically designed for a wide range of bacterial species. However, differences in cell wall composition amongst bacteria can compromise the efficiency of RNA isolation with these kits, particularly when working with non-model bacteria like *Fusobacterium nucleatum*. In addition, column-based purification methods often

Hung Ton-That (ed.), *Fusobacterium nucleatum: Methods and Protocols*, Methods in Molecular Biology, vol. 3055,
https://doi.org/10.1007/978-1-0716-5388-3_10,

impose size restriction on extracted transcripts, thereby introducing bias and making it difficult to study gene expression changes of shorter transcripts, such as small regulatory RNAs (sRNAs).

Here, we present a protocol for isolating high yields of *F. nucleatum* RNA that is free of DNA or other contaminants. The described steps below include the efficient lysis of *F. nucleatum* cells collected from liquid culture, the extraction of nucleic acids, and removal of genomic DNA. The combination of the steps enables the generation of high-quality RNA-seq libraries.

By applying this method across all *Fusobacterium* subspecies as well as the closely related *F. periodonticum*, we were able to characterize the transcriptomes of these subspecies to precisely map transcriptional start sites and transcriptome structure [1]. Importantly, because our approach retains small transcripts, it also facilitated the discovery of small noncoding RNAs (sRNAs) in *Fusobacterium* [1–3], a class of transcripts that play an increasingly recognized role in regulating a bacterium's adaption processes. This protocol can be used in future studies directed at exploring transcriptional responses of *F. nucleatum* to varying environmental conditions.

2 Materials

To minimize RNA degradation, all water used during the experiments, including that used for buffer preparations, should be treated with DEPC to deactivate RNases. All steps should be performed in safe-lock tubes to reduce the risk of exposure to organic solvents like phenol or chloroform during mixing steps (*see* **Note 1**).

2.1 Sample Collection and Bacterial Lysis

1. STOP-mix: 95% of 100% molecular biology grade ethanol and 5% of phenol.
2. Liquid nitrogen.
3. −80 °C freezer.
4. Refrigerated centrifuge capable of ~4500 × *g*.
5. Tris-EDTA buffer, pH 8.
6. 10% sodium dodecyl sulfate (SDS).
7. Water bath/dry bath (64 °C).
8. 3 M of sodium acetate, pH 5.2.
9. 50 μg/mL of lysozyme.

2.2 Phase Separation Steps

1. Acidic phenol.
2. Chloroform.
3. Phase-lock gel (PLG) tubes (optional).

4. phenol:chloroform:isoamyl alcohol (P/C/I, 25:24:1 (v/v)).
5. Water bath/dry bath (64 °C).
6. Refrigerated centrifuge capable of ~13,000 × *g*.

2.3 Nucleic Acid Precipitation

1. −20 °C freezer.
2. Precipitation mix: 29 parts: 100% molecular biology grade ethanol, 1 part: 3 M of sodium acetate, pH 6.5.
3. GlycoBlue (optional).
4. Refrigerated centrifuge capable of ~13,000 × *g*.
5. 75% molecular biology grade ethanol.
6. DEPC water.
7. Thermoshaker or shaker (1000 rpm).

2.4 DNA Removal

1. RNase-free Dnase I (2000 U).
2. 10× Dnase I buffer.
3. Rnase inhibitor (0.5 U).
4. Heat block or incubator (37 °C).
5. DEPC water.
6. Nanodrop or Qubit for quantification of nucleic acids.

3 Methods

3.1 Collection and Fixation of Bacterial Cells

1. Harvest bacteria at the desired time point. The total optical density of the collected culture should not exceed a combined OD_{600} of 6.
2. Add 0.2 volumes of ice-cold STOP mix to the collected sample.
3. Invert samples for mixing and freeze samples in liquid nitrogen.
4. Store at −80 °C until further processing.

3.2 Sample Preparation

1. Thaw samples on ice.
2. Centrifuge at 4500 × *g* for 10 min at 4 °C.
3. Decant the supernatant and retain the pellet.
4. Briefly spin down samples at 4500 × *g* for 30 s at 4 °C.
5. Remove remaining supernatant with pipette to avoid carryover of phenol.

3.3 Cell Lysis and First Extraction

1. Resuspend the pellet in 600 μL of Tris-EDTA buffer containing 50 μg/mL lysozyme (*see* **Note 2**).
2. Add 60 μL of 10% SDS and mix by inverting the tubes.

3. Incubate at 64 °C for 2 min.
4. Place tubes back on ice.
5. Add 66 μL of 3 M of sodium acetate (pH 5.2), and invert tubes gently. A white precipitate should be visible.
6. Add 750 μL of acidic phenol and mix vigorously for at least 10 s.
7. Centrifuge at 13,000 × *g* for 15 min at 4 °C to separate phases.

3.4 Second Extraction and Nucleic Acid Precipitation

1. Transfer the aqueous phase to phase-lock gel (PLG) tubes (*see* **Note 3**).
2. Add 750 μL of chloroform, and mix vigorously.
3. Centrifuge again at 13,000 × *g* for 15 min at 4 °C.
4. Transfer the aqueous phase to a fresh 2-mL tube.
5. Add 1.4 mL of precipitation mix to the aqueous phase.
6. Optional: Add 1 μL of GlycoBlue to support precipitation and visualization of the nucleic acid pellet.
7. Incubate at −20 °C overnight to precipitate nucleic acids (*see* **Note 4**).

3.5 RNA Recovery

1. Centrifuge at 13,000 × *g* for 30 min at 4 °C.
2. Discard the supernatant and wash the pellet with 75% ethanol.
3. Briefly spin down at 13,000 × *g* for 30 s at 4 °C.
4. Carefully remove remaining supernatant with pipette without disturbing the pellet.
5. Dry the pellet at room temperature for 15 min.
6. Resuspend the nucleic acid pellet in 80 μL of RNase-free water (*see* **Note 5**).
7. Incubate at 65 °C with shaking at 1000 rpm for 5 min.
8. Store at −20 °C or proceed to DNase treatment (*see* **Note 6**).

3.6 Removal of DNA and Post-Digestion Purification

1. Determine the nucleic acid concentration of each sample.
2. Prepare 40 μg of extracted nucleic acids as input for the next step in fresh tubes. Adjust the volume to a total volume of 30 μL.
3. Prepare a DNA digestion master mix containing the following per sample: 9.5 μL of RNase-free water, 5 μL of 10xDNase I buffer, 5 μL of RNase-free DNase, and 0.5 μL of RNase inhibitor. Calculate the amounts required for the master mix with one extra sample to account for pipetting errors.
4. Add 20 μL of DNase digestion master mix to each sample and mix.
5. Incubate the samples at 37 °C for 1 h.

6. Add 2 volumes of RNase-free water to the DNase-treated mix.
7. Transfer to PLG tubes, and add an equal volume of phenol: chloroform:isoamyl alcohol (P/C/I).
8. Mix vigorously and centrifuge at 13,000 × *g* for 15 min at 4 °C.

3.7 Final RNA Precipitation and Cleanup

1. Transfer the aqueous phase to a clean tube.
2. Add two volumes of precipitation mix.
3. Optional: Add 1 μL of GlycoBlue to support precipitation and visualization of the nucleic acid pellet.
4. Precipitate overnight at −20 °C (*see* **Note 4**).
5. Centrifuge at 13,000 × *g* for 30 min at 4 °C.
6. Discard the supernatant and wash the pellet with 75% ethanol.
7. Briefly spin down at 13,000 × *g* for 30 s at 4 °C.
8. Carefully remove remaining supernatant with pipette without disturbing the pellet.
9. Dry the pellet at room temperature for 15 min.
10. Resuspend the RNA pellet in 30 μL of RNase-free water (*see* **Note** 7).
11. Incubate at 65 °C with shaking at 1000 rpm for 5 min.
12. Resulting RNA is DNA-free and ready for downstream applications.

4 Notes

1. All samples are to be kept on ice if not otherwise specified. The use of metal cooling racks is recommended to ensure stable temperatures and easier sample handling.
2. Ensure that the entire pellet is completely resuspended. Any remaining clumps will diminish the lysis.
3. The use of PLG tubes is recommended but not strictly required. Careful removal of the aqueous phase in the individual steps yields similarly high-quality RNA but may reduce yield and increases the time needed during the steps.
4. This step is a good stopping point. The samples can be kept at −20 °C for 48–72 h. Alternatively, the precipitation time can be reduced by the following two approaches. Incubation of the samples at −80 °C for ~1 h or directly moving on to **step 6**. For the latter, the centrifugation time should be extended to 90–120 min.
5. The initial harvesting step can be scaled up based on experimental needs. If scaled, the volume of the RNase-free water in

step 1 should be adjusted proportionally to match the increased input material.

6. While direct processing is recommended, the samples can be stored at this step for ~2 weeks at −20 °C or ~6 months at −80 °C if needed.
7. The volume of the RNase-free water should be adjusted to yield the desired RNA concentration according to the need for the application. The current protocol should yield ~1 μg/μL.

Conflict of Interest

The authors have no conflicts of interest to declare that are relevant to the content of this chapter

References

1. Ponath F, Tawk C, Zhu Y, Barquist L, Faber F, Vogel J (2021) RNA landscape of the emerging cancer-associated microbe Fusobacterium nucleatum. Nat Microbiol 6(8):1007–1020. https://doi.org/10.1038/s41564-021-00927-7
2. Ponath F, Zhu Y, Cosi V, Vogel J (2022) Expanding the genetic toolkit helps dissect a global stress response in the early-branching species Fusobacterium nucleatum. Proc Natl Acad Sci USA 119(40):e2201460119. https://doi.org/10.1073/pnas.2201460119
3. Ponath F, Zhu Y, Vogel J (2024) Transcriptome fine-mapping in Fusobacterium nucleatum reveals FoxJ, a new sigma(E)-dependent small RNA with unusual mRNA activation activity. MBio 15(4):e0353623. https://doi.org/10.1128/mbio.03536-23

Chapter 11

Isolation and Visualization *Fusobacterium nucleatum* Microcompartments

Dana Franklin, Angela Agnew, John Crow, Z. Hong Zhou, and Hung Ton-That

Abstract

In *Fusobacterium nucleatum*, bacterial microcompartments (BMCs) are polyhedral organelles that encapsulate the catabolism of ethanolamine (EA), creating a specialized environment that protects the rest of the cell from toxic metabolic byproducts. In this chapter, we present a protocol for the isolation and visualization of *F. nucleatum* BMCs with transmission electron microscopy (TEM), cryogenic electron microscopy (cryoEM), and cryogenic electron tomography (cryoET). The protocol includes anaerobic cultivation of *F. nucleatum*, a strategy for BMC purification, sample preparation and image acquisition for TEM, cryoEM, and cryoET, followed by 3D reconstruction and visualization. These approaches are adaptable to BMCs from other bacterial species, offering a robust framework for structural and functional studies of diverse BMC systems.

Key words *Fusobacterium nucleatum*, Ethanolamine, Bacterial microcompartments, Electron microscopy, Cryogenic electron microscopy

1 Introduction

Electron microscopy (EM) is a fundamental tool for visualizing subcellular structures in biological samples, offering direct insights into the size, morphology, and organization of macromolecular complexes and organelle-like structures such as bacterial microcompartments (BMCs). Among EM techniques, transmission electron microscopy (TEM) stands out as a conventional and rapid method for assessing negatively stained or thin-sectioned samples. TEM enables straightforward evaluation of the integrity, size, and overall morphology of isolated BMCs [1–3], making it an accessible approach for preliminary sample quality assessment prior to more

---

Dana Franklin and Angela Agnew contributed equally with all other contributors.

Hung Ton-That (ed.), *Fusobacterium nucleatum: Methods and Protocols*, Methods in Molecular Biology, vol. 3055, https://doi.org/10.1007/978-1-0716-5388-3_11,

advanced structural analysis sample quality prior to more advanced structural analyses [4].

Cryogenic electron microscopy (cryoEM) has emerged as a powerful and rapidly advancing technique for nanoscale examination and structural determination of frozen-hydrated biological specimens. It facilitates three-dimensional (3D) reconstruction of structurally homogeneous complexes at near-atomic resolution, without requiring extensive purification or crystallization. CryoEM can visualize particles ranging from single proteins to large complexes containing tens of subunits, typically those exceeding 50 kDa [5]. For pleomorphic structures such as BMCs or entire frozen cells, cryogenic electron tomography (cryoET) enables 3D reconstruction by capturing multiple viewing angles through stage tilting [6–9]. Samples are flash-frozen in liquid ethane, vitrifying the water content and preserving proteins in their native state, thereby eliminating the need for heavy fixatives and stains [10]. Together, these EM modalities provide complementary capabilities for both rapid screening and high-resolution structural analysis of polydisperse BMC populations found in various bacterial species, including *Fusobacterium nucleatum*.

F. nucleatum is a Gram-negative, obligate anaerobe predominantly residing in the human oral cavity. Beyond its oral niche, *F. nucleatum* can enter the bloodstream and disseminate to distal organs such as the colon and placenta. It has been implicated in periodontal disease and a range of extraoral pathologies, including adverse pregnancy outcomes and colorectal cancer [11, 12]. Like many bacteria, *F. nucleatum* utilizes BMCs to optimize metabolic pathways and facilitate processes related to pathogenesis [13]. Notably, BMC production in *F. nucleatum* increases in response to ethanolamine exposure [3]. The organism harbors a 21-gene cluster encoding shell proteins that self-assemble into polyhedral organelles. These BMCs mediate ethanolamine catabolism while sequestering toxic intermediates from the cytoplasm [3].

Here we present a protocol for cryogenic grid preparation and visualization of *F. nucleatum* BMCs using TEM, cryoEM, and cryoET (*see* Figs. 1 and 2). Room-temperature TEM provides high-contrast, low-resolution two-dimensional (2D) spatial information regarding sample concentration and condition (*see* Fig. 1b). CryoEM yields noisy but high-resolution 2D images and, when sufficient copies of the target are present, enables high-resolution 3D reconstructions (*see* Fig. 1c). CryoET, through stage tilting, captures multiple angles of a single viewing area, producing low-resolution 3D reconstructions of unique, non-repeating samples (*see* Fig. 2).

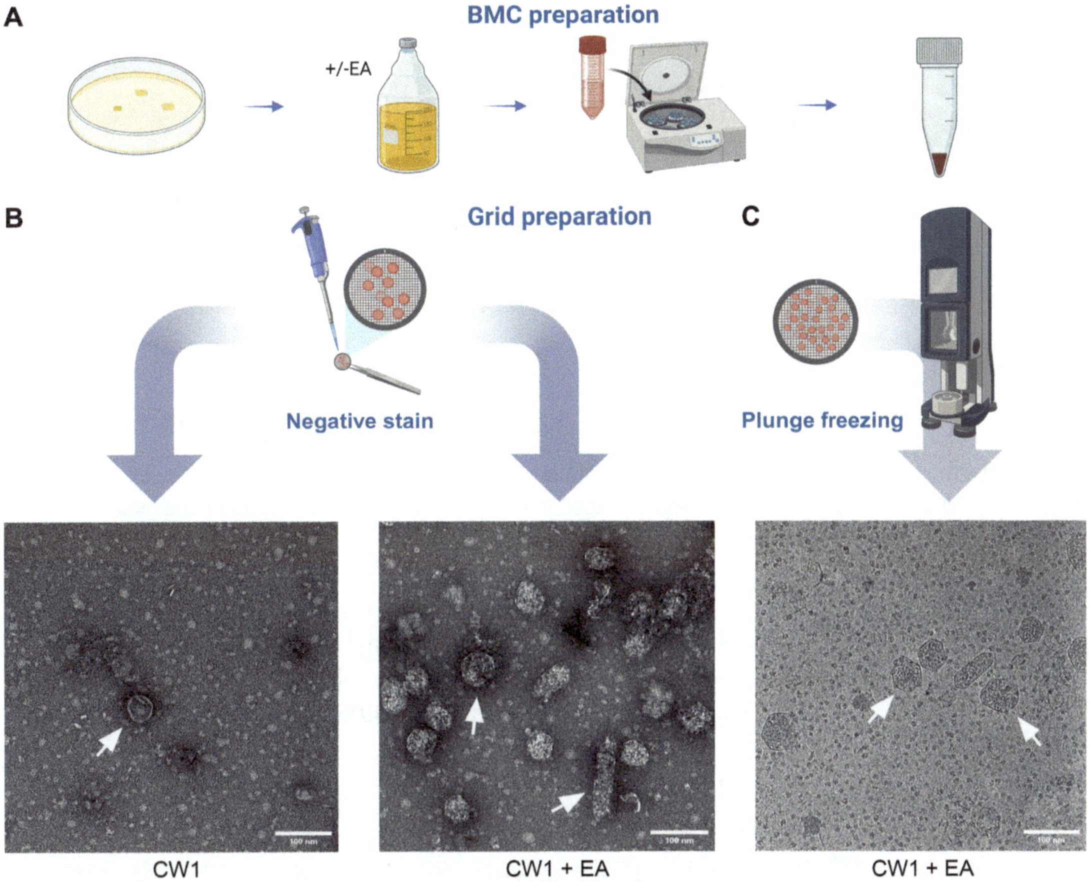

Fig. 1 Transmission and cryogenic electron microscopy of bacterial microcompartments (BMCs). (**a**) Workflow overview of cell culture and BMC isolation. Colonies are isolated and cultured anaerobically with or without ethanolamine, followed by chemical lysis and differential centrifugation for separation and concentration of BMCs. (**b**) BMCs were isolated from *F. nucleatum* CW1 (a derivative of strain ATCC 23726) grown in the presence or absence of ethanolamine (EA). Samples were negatively stained with 2% uranyl acetate (top) and visualized by transmission electron microscopy (bottom) to determine sample concentration, general morphology, and quality; scale bars of 100 nm. (**c**) BMCs were plunge-frozen (top) and imaged via cryoEM, with their fully hydrated, polyhedral morphologies visible (bottom); scale bars of 100 nm

2 Materials

2.1 Cultivation of Bacterium and Isolation of BMCs

1. *F. nucleatum* ATCC 23726.
2. Anaerobic chamber (10% CO_2, 10% H_2, and 80% N_2).
3. TSPC: Tryptic Soy Broth (TSB) supplemented with 1% Bacto peptone and 0.05% autoclaved cysteine.
4. TSPC agar: Tryptic Soy Broth (TSB) supplemented with 1% Bacto. peptone, 1.5% Agar, and 0.05% autoclaved cysteine.
5. 1% cysteine.

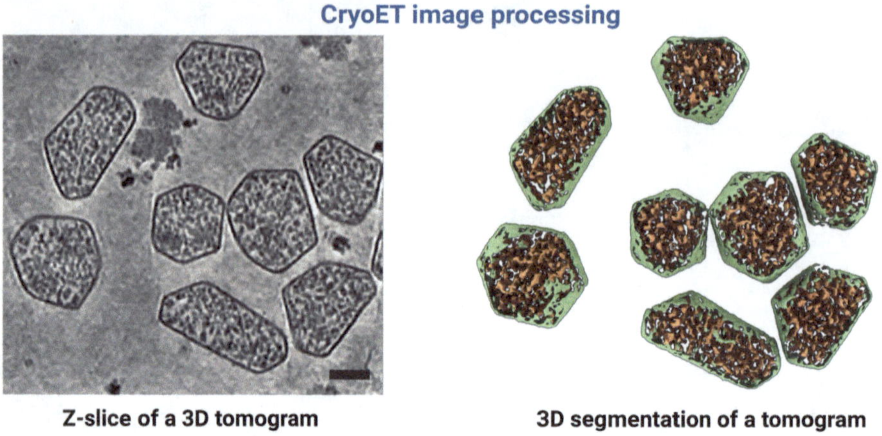

Fig. 2 An example of a tomogram reconstructed from cryoET data collection. These were visualized via Z-slice (left) and by rendering in ChimeraX with segmentation to differentiate shell protein (green) and interior proteins (orange). Scale bar, 50 nm

6. 5× M9 salts: 64 g of $Na_2HPO_4 \bullet 7H_2O$, 15 g of KH_2PO_4, 2.5 g of NaCl, and 5 g of NH_4Cl in 1 L of Milli-Q water.
7. Minimal media (MM): 5× M9 Salt, 1% Bacto peptone, 0.5 M $MgSO_4$, 0.1 mM $CaCl_2$, and 0.05% autoclaved cysteine in DI water.
8. 5 M ethanolamine (EA).
9. 1 M PMSF.
10. 1 mg/mL of lysozyme.
11. 10 μg/mL of DNAse I.
12. Bacterial protein extraction reagent (BPER).
13. Buffer A: 50 mM of Tris-HCl, 500 mM of KCl, and 12.5 mM of $MgSO_4$.
14. Buffer B: 50 mM of Tris-HCl, 50 mM of KCl, and 5 mM of $MgCl_2$.
15. Milli-Q water.
16. Deionized water (DI water).

2.2 Transmission Electron Microscopy

1. *F. nucleatum* purified BMCs.
2. Formar/carbon film 200 mesh, nickel grid.
3. Dumont tweezer N4AC.
4. 2% uranyl acetate solution.
5. Kimwipes.
6. 200 kV TF20 high-resolution EM.

2.3 CryoEM and CryoET

1. *F. nucleatum* purified BMCs.
2. Vitrobot filter paper, Grade 595.
3. Fine-tipped tweezers.
4. Lacey carbon grids.
5. PELCO easiGlow™ glow discharge cleaning system.
6. Vitrobot Mark IV cryo-sample plunger.
7. 200 kV TF20 electron microscope.
8. 300 kV Titan Krios electron microscope.

3 Methods

Prepare all solutions using ultrapure MQ water except for minimal media solutions.

3.1 Cultivation of Bacterium and Isolation of BMCs

1. Streak out *F. nucleatum* ATCC 23726 cells from −80 °C freezer on TSPC agar plates.
2. Allow to grow for 2 days in anaerobic chamber at 37 °C.
3. Inoculate a single colony from a plate in 5 mL of TSPC, and grow overnight (~14–16 h) in the anaerobic chamber at 37 °C.
4. Refresh by adding ~800 μL of overnight culture to 8 mL TSPC, and grow to OD_{600} ~0.6–0.8 (6–8 h).
5. Prepare 800 mL MM with 0.05% cysteine (*see* **Note 1**).
6. Place respective cultures in 800 mL MM (pH 6.9) with or without EA.
7. After 14–16 h, harvest fusobacterial cells by centrifugation at 5000 × g for 15 min at room temperature.
8. Wash cells in 20 mL of Buffer A, and spin down at 5000 × g for 15 min at room temperature.
9. Resuspend bacterial pellets in 10 mL of 60% BPER in Buffer A with DNase I (0.08 mg/mL), 50 mg lysozyme, and 1 mM PMSF.
10. Rock in a 30 °C shaker at 150 rpm for 1 h. Remaining steps should be performed at 4 °C to maintain BMC stability.
11. Centrifuge at 12,000 × g for 8 min.
12. Transfer the supernatant to a new 15-mL conical tube, and centrifuge at 20,000 × g for 40 min.
13. Resuspend crude BMCs in 50–125 μL of chilled Buffer B, and transfer to an Eppendorf microcentrifuge tube (*see* **Note 2**).
14. Spin the microcentrifuge tube down at 12,000 × g for 5 min.

15. Transfer the supernatant, which contains isolate BMCs, to a new microcentrifuge tube.
16. Right after BMC extraction, prepare EM grids following the procedure below (*see* **Note 3**).

3.2 Transmission Electron Microscopy

1. Place 7 μL of isolated BMCs on the carbon/nickel grid, and let stand for ~3 min.
2. Wick out the liquid with Kimwipe, and add 7 μL of 2% uranyl acetate for 2 min.
3. After 2 min, wick out the liquid, and immediately add another 7 μL of 2% uranyl acetate.
4. Immediately wick out the liquid, making sure that the grid does not become dry.
5. Turn tweezers 90° and allow to dry completely.
6. Store prepared grids in a grid box for future electron microscopy.
7. Analyze and image grids using an electron microscope (TF20) at a nominal magnification of 29,000× with a room temperature single tilt holder.

3.3 CryoEM and CryoET

3.3.1 Grid Freezing

1. Prepare Vitrobot chamber with conditions 4 °C and 100% humidity.
2. Set blotting parameters accordingly per Vitrobot specifications; for BMCs on lacey carbon grids, a longer blot time and a wait time of at least 5 s is recommended. Allow the saturation of blotting paper for 20 min before use.
3. Place fresh lacey carbon grids on clean filter paper in a glass petri dish, and glow discharge for 30 s with a target current of 15 mA.
4. Prepare cryogen from a 30%/70% ethane-propane mixture.
5. Secure a glow-discharged grid in tweezers, and mount on Vitrobot such that the grid is pointing downward.
6. Raise tweezers and cryogen; add 3 μL of samples to the grid through the side entry port of the Vitrobot.
7. Plunge the grid with samples into the cryogen, and wait for 30 s before transferring to a cryo-storage box under liquid nitrogen (*see* **Note 4**).
8. Screen cryoEM grids on a TF20 microscope for ice thickness and sample quality with a cryo-holder.
9. Iteratively adjust freezing parameters or concentration until BMCs fill the majority of the field of view and the ice is thin (e.g., grid squares at low magnification appear to have soft corners, contrast of particles is good at high magnification).

10. Evaluate particle quality at 29000× nominal magnification with low-dose settings.
    (i) If ice is too thick, increase blotting time and blotting force (*see* **Note 5**).
    (ii) If particle concentration is too low, resuspend BMCs in a smaller volume at the end of purification.
11. Once appropriate freezing conditions are determined, freeze 8–12 grids of the same condition, and store in liquid nitrogen until a suitable date of data collection.

3.3.2 Data Collection

1. Load the saved grids onto a 300 kV Titan Krios microscope equipped with a Falcon 4i direct electron detector with the Tomo5 software [14]. Align and set acquisition parameters in the ThermoFisher Scientific Tomography 5 software (*see* **Note 6**).
2. Acquire a low-magnification atlas of each grid at 125×, and then pick squares for data collection. At each selected square, create a medium (3600×) magnification map to identify and select regions with targets visible. Only consider regions with appropriate ice thickness (enough contrast to see BMCs at medium magnification).
3. Collecting final tilt series images at 105,000× will give rise to a pixel size of 1.24 Å/px. In cryoET it is typical to set tilting from −60° to 60° in increments of 3°, with an exposure of around 3 electrons/$Å^2$ per tilt and an accumulated electron dose of 120 electrons/$Å^2$ over the whole tilt stack.
4. Tilt series are processed using the Relion5 tomo software package for motion correction, tilt series alignment, and tomogram reconstruction [15]. Correct for missing-wedge-induced anisotropy using the IsoNet software package for improved map interpretability [8]. Visualization and map segmentation can be performed in ChimeraX [16].

4 Notes

1. When making the minimal media culture, it is best to use autoclaved deionized water.
2. The final suspension number of BMCs in Buffer B is dependent on pellet size. Ensure that you are resuspending the BMCs gently.
3. Due to the instability of BMCs, it is best to visualize them within 24 h to ensure reliable results.
4. Best practice calls for minimal exposure of liquid nitrogen and cryogen to ambient humidity to avoid condensation of

crystalline ice from the atmosphere. Keep the cryogen and clean nitrogen dewars covered.

5. Adjustments to sample concentration, sample purity, grid type, and filter paper grade will inevitably change the necessary blotting conditions and require iterative screening.
6. Instrument specifications and data collection software will depend upon the resource availability at your cryoEM facility. Stated magnifications will vary depending upon the instrument. Consult with facility employees for settings of best performance.

Acknowledgments

Research reported in this publication was supported by the National Institutes of Health (R21DE032906 to HTT and R01GM071940 to ZHZ). We acknowledge the use of resources at the Electron Imaging Center for Nanosystems supported by NIH (S10OD032459) and NSF (DMR-1548924). Dana Franklin was supported by the Ruth L. Kirschstein National Research Service Award T32AI007323. Angela Agnew was supported by the UCLA Dentist-Scientist and Oral Health-Researcher Training Program, NIDCR grant T90DE030860.

Conflict of Interest The authors have no conflicts of interest to declare that are relevant to the content of this chapter.

References

1. Yeates TO, Kerfeld CA, Heinhorst S, Cannon GC, Shively JM (2008) Protein-based organelles in bacteria: carboxysomes and related microcompartments. Nat Rev Microbiol 6(9): 681–691. https://doi.org/10.1038/nrmicro1913
2. Tanaka S, Kerfeld CA, Sawaya MR, Cai F, Heinhorst S, Cannon GC, Yeates TO (2008) Atomic-level models of the bacterial carboxysome shell. Science 319(5866):1083–1086. https://doi.org/10.1126/science.1151458
3. Franklin DS, Chen YW, Chen Y, Wittchen M, Agnew A, Luu A, Whitelegge JP, Zhou ZH, Tauch A, Das A, Ton-That H (2025) Ethanolamine-induced assembly of microcompartments is required for Fusobacterium nucleatum virulence. MBio 16(2):e0340524. https://doi.org/10.1128/mbio.03405-24
4. Bharat TA, Scheres SH (2016) Resolving macromolecular structures from electron cryo-tomography data using subtomogram averaging in RELION. Nat Protoc 11(11): 2054–2065. https://doi.org/10.1038/nprot.2016.124
5. Zhou ZH (2008) Towards atomic resolution structural determination by single-particle cryo-electron microscopy. Curr Opin Struct Biol 18(2):218–228
6. Nogales E, Mahamid J (2024) Bridging structural and cell biology with cryo-electron microscopy. Nature 628(8006):47–56. https://doi.org/10.1038/s41586-024-07198-2
7. Vilas JL, Carazo JM, Sorzano COS (2022) Emerging themes in CryoEM horizontal line single particle analysis image processing. Chem Rev 122(17):13915–13951. https://doi.org/10.1021/acs.chemrev.1c00850
8. Liu Y-T, Zhang H, Wang H, Tao C-L, Bi G-Q, Zhou ZH (2022) Isotropic reconstruction for

electron tomography with deep learning. Nat Commun 13(1):6482. https://doi.org/10.1038/s41467-022-33957-8

9. Wang H, Liao S, Yu X, Zhang J, Zhou ZH (2024) TomoNet: a streamlined cryogenic electron tomography software pipeline with automatic particle picking on flexible lattices. Biological Imaging 4:e7. https://doi.org/10.1017/S2633903X24000060
10. Weissenberger G, Henderikx RJM, Peters PJ (2021) Understanding the invisible hands of sample preparation for cryo-EM. Nat Methods 18(5):463–471. https://doi.org/10.1038/s41592-021-01130-6
11. Han YW, Redline RW, Li M, Yin L, Hill GB, McCormick TS (2004) Fusobacterium nucleatum induces premature and term stillbirths in pregnant mice: implication of oral bacteria in preterm birth. Infect Immun 72(4):2272–2279
12. Rubinstein MR, Wang X, Liu W, Hao Y, Cai G, Han YW (2013) Fusobacterium nucleatum promotes colorectal carcinogenesis by modulating E-cadherin/beta-catenin signaling via its FadA adhesin. Cell Host Microbe 14(2):195–206. https://doi.org/10.1016/j.chom.2013.07.012
13. Bobik TA, Lehman BP, Yeates TO (2015) Bacterial microcompartments: widespread prokaryotic organelles for isolation and optimization of metabolic pathways. Mol Microbiol 98(2):193–207. https://doi.org/10.1111/mmi.13117
14. Scientific TF (2025) Tomography 5 software, v5.6. Thermo Fisher Scientific
15. Burt A, Toader B, Warshamanage R, von Kugelgen A, Pyle E, Zivanov J, Kimanius D, Bharat TAM, Scheres SHW (2024) An image processing pipeline for electron cryo-tomography in RELION-5. FEBS Open Bio 14(11):1788–1804. https://doi.org/10.1002/2211-5463.13873
16. Meng EC, Goddard TD, Pettersen EF, Couch GS, Pearson ZJ, Morris JH, Ferrin TE (2023) UCSF ChimeraX: tools for structure building and analysis. Protein Sci 32(11):e4792. https://doi.org/10.1002/pro.4792

Chapter 12

Isolation and Proteomic Characterization of Outer Membrane Vesicles in *Fusobacterium nucleatum*

Jinjing Liu and Yung-Fu Chang

Abstract

Fusobacterium nucleatum is an anaerobic Gram-negative bacterium associated with periodontal disease and colorectal cancer. Outer membrane vesicles (OMVs) produced by *F. nucleatum* are nanoscale lipid vesicles enriched with immunogenic and virulence-associated proteins, making them promising candidates for vaccine development. This protocol describes a comprehensive workflow for culturing *F. nucleatum*, isolating and purifying OMVs via ultracentrifugation and density gradient ultracentrifugation, and characterizing their biophysical properties using transmission electron microscopy (TEM) and dynamic light scattering (DLS). Proteomic profiling of purified OMVs was conducted using LC-MS/MD, and the resulting data were analyzed through bioinformatics tools. Protein function, subcellular localization, and immunogenic features were predicted using Vaxign, PRED-TMBB2, InterPro, and EMBOSS. Pathway and functional classification were performed using KEGG and EggNOG databases. Together, this methodology provides a reproducible pipeline for characterizing OMV-associated proteins and identifying potential antigens for *F. nucleatum*.

Key words Outer membrane vesicles, *Fusobacterium nucleatum*, Proteomics, Vaccine candidates

1 Introduction

Outer membrane vesicles (OMVs) are spherical, bilayer nanoparticles that are naturally secreted by Gram-negative bacteria [1]. OMVs carry diverse cargo, including proteins, lipids, and nucleic acids. Due to their intrinsic immunogenicity and ability to present bacterial surface antigens in a non-replicative form, OMVs are increasingly recognized as promising platforms for vaccine development [2, 3].

Fusobacterium nucleatum is an obligate anaerobic Gram-negative bacterium commonly found in the human oral cavity and gastrointestinal tract [4–6]. It plays a central role in periodontal disease and has been found strongly associated with colorectal cancer (CRC) [7–9]. Recent studies have identified *F. nucleatum* as a carrier of virulence factors, including adhesins such as FadA and

Hung Ton-That (ed.), *Fusobacterium nucleatum: Methods and Protocols*, Methods in Molecular Biology, vol. 3055, https://doi.org/10.1007/978-1-0716-5388-3_12,

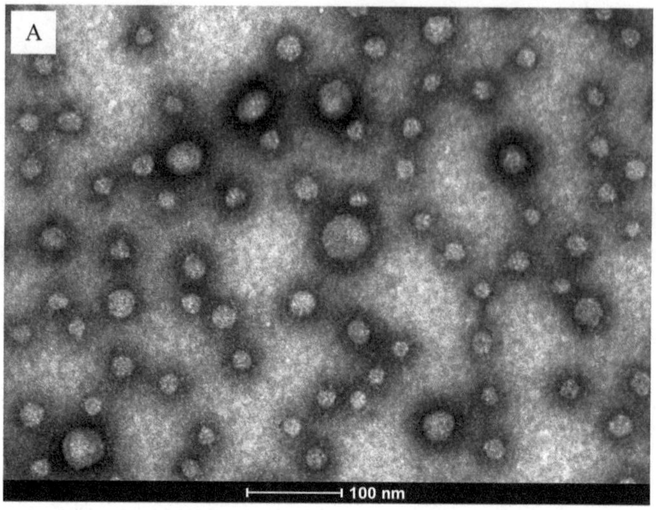

Fig. 1 Morphology of purified outer membrane vesicles (OMVs) from *F. nucleatum*. Negatively stained OMVs were visualized using transmission electron microscopy; a scale bar of 100 nm

Fap2, suggesting that OMV-based strategies are useful for studying pathogenesis and developing immune interventions.

Here, we present a detailed protocol for the isolation, purification, and proteomic analysis of OMVs from *F. nucleatum* subsp. *animalis 7_1*. This workflow includes anaerobic bacterial culture, collection, and density gradient purification of OMV, as well as biophysical characterization using transmission electron microscopy (TEM; *see* Fig. 1) and dynamic light scattering (DLS, *see* Fig. 2). Additionally, we outline a comprehensive mass spectrometry-based proteomics pipeline, combined with bioinformatic analyses to annotate protein localization, function, and immunogenic potential [10]. This protocol provides a reliable approach for profiling OMVs from *F. nucleatum*.

2 Materials

2.1 Bacterial Culture

1. *F. nucleatum* subsp. *Animalis* 7_1 (EAVG_002).
2. Tryptic soy broth (TSB) supplemented with 5 μg/mL of hemin and 5 μg/mL of menadione.
3. Ultrapure water (Mili-Q or equivalent).
4. Fastidious Anaerobe Agar.
5. Horse blood.
6. 50% (v/v) glyrerol in water.
7. Spectrometer.
8. Cryogenic tubes.

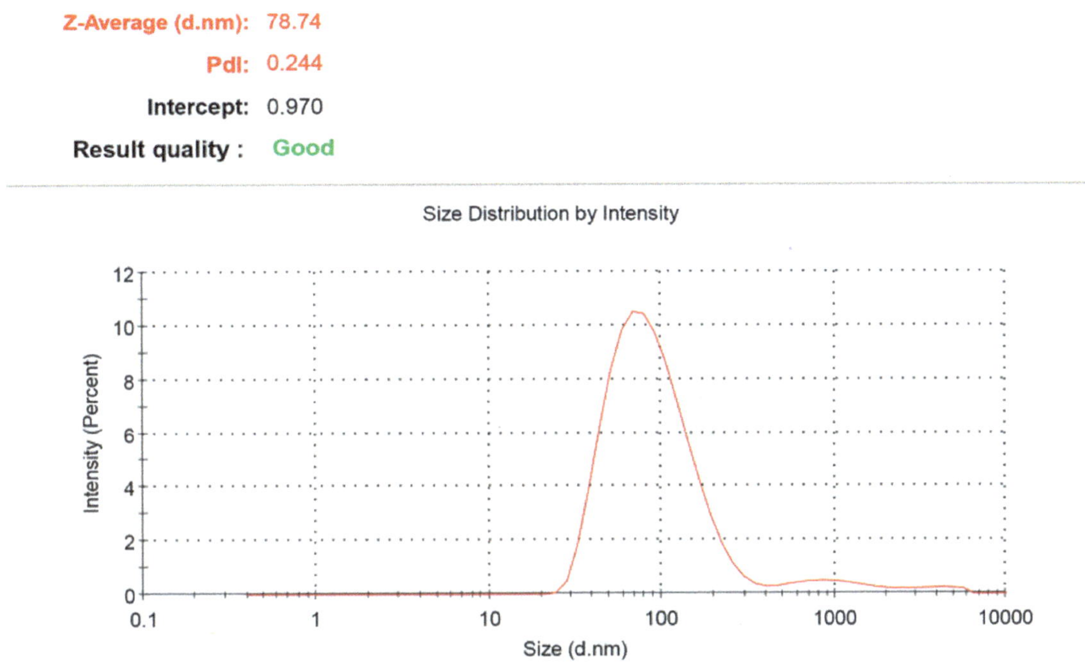

Fig. 2 Dynamic light scattering analysis. Shown is the dynamic light scattering analysis of *F. nucleatum* outer membrane vesicles

2.2 OMV Isolation and Purification

1. 0.22-μm vacuum filter membranes.
2. OptiPrep density gradient media (60% iodixanol; Sigma-Aldrich or equivalent).
3. Tricine-NaOH buffer: 10 mM, pH 7.4, and 0.85% NaCl.
4. Ultracentrifuge (Beckman-Coulter or equivalent).
5. Type 45 Ti rotors, SW41 Ti rotor (Beckman-Coulter).
6. Polypropylene ultracentrifuge tubes.
7. Steriled phosphate-buffered saline (PBS).

2.3 Transmission Electron Microscopy

1. 1.5% uranyl acetate.
2. 200-mesh carbon-coated copper grids.
3. Electron microscope (Thermo-Fisher Tecnai Spirit Twin STEM or equivalent).

2.4 Dynamic Light Scattering (DLS)

1. DLS equipment (Malvern Zetasizer Nano ZS DLS or equivalent).

2.5 SDS-PAGE and In-Gel Digestion

1. 12% SDS-PAGE.
2. Coomassie Brilliant Blue staining solution: 0.1% Coomassie Brilliant Blue R-250, 50% (v/v) methanol, and 10% (v/v) acetic acid in water.

3. Gel electrophoresis equipment.
4. Destaining solution: 50% (v/v) methanol and 10% (v/v) acetic acid in water.
5. Scalpel.
6. 100-mM ammonium bicarbonate.
7. Acetonitrile.
8. Reduction buffer: 10 mM of dithiothreitol (DTT) in 100 mM of ammonium bicarbonate.
9. SpeedVac or a vacuum concentrator.
10. 37 °C incubator.
11. Alkylation buffer: 55 mM of iodoacetamide in 100 mM of ammonium bicarbonate.
12. Trypsin digestion buffer: 20 ng/mL of sequencing-grade modified trypsin in 50-mM ammonium bicarbonate with 10% (v/v) acetonitrile.
13. Peptide extraction solution 1: 50% (v/v) acetonitrile and 5% (v/v) formic acid in water.
14. Peptide extraction solution 2: 90% (v/v) acetonitrile and 5% (v/v) formic acid in water.
15. Final resuspension solution: 0.5% (v/v) formic acid in water.

2.6 LC-MS/MS Analysis

1. Orbitrap fusion tribrid mass spectrometer (Thermo Fisher Scientific or equivalent).
2. Nano-liquid chromatography system (UltiMate 3000 RSLCnano system from Thermo Fisher Scientific or equivalent).
3. Trap column: C18 reverse-phase, 5-μm particle size, and 100-μm inner diameter × 20 mm.
4. Analytical column: C18 reverse-phase, 2-μm particle size, and 75-μm inner diameter × 25 cm.
5. Nanospary ionization source (Nanospray Flex from Thermo Fisher Scientific or equivalent).
6. Temperature-controlled column oven, compatible with the chromatography system.
7. Autoampler vials (LC-MS grade glass).
8. Low-rentention microcentrifuge tubes: 1.5-mL polypropylene.
9. LC-MS grade solvents: acetonitrile, formic acid, and water.
10. Vacuum concentrator or lyophilizer.
11. Xcalibur software (Thermo Fisher Scientific).

2.7 Data Analysis and Protein Annotation

1. Proteome Discoverer 2.2 with the Sequest HT algorithm.
2. Bioinformatics tools: InterProScan, LipoP 1.0, PRED-TMMB2, CELLO2go, Vaxign, KEGG (Kyoto Encyclopedia of Genes and Genomes), EggNOG, EMBOSS Antigenic, BCPREDS, Vaxitop, and R software with packages: ggplot2.

3 Methods

3.1 Bacterial Culture and Harvest

1. Streak *F. nucleatum* on Fastidious anaerobe agar plates supplemented with 5% defibrinated horse blood, and incubate them under anaerobic conditions (80% N_2, 10% CO_2, 10% H2) at 37 °C [8].
2. Inoculate a single colony into 200-mL TSB supplemented with 5 μg/mL of hemin and 5 μg/mL of menadione, and grow anaerobically to an optical density at 600 nm (OD_{600}) = 0.7 ~ 0.9 (*see* **Note 1**).
3. Store these cultures with an equal volume of 50% glycerol at −80 °C in cryogenic storage tubes for long-term use.

3.2 OMV Isolation and Purification

1. Centrifuge 200 mL of bacterial culture at 8500× *g* for 15 min at 4 °C.
2. Filter the supernatant through a 0.22-μm membrane fillter.
3. Ultracentrifuge the filtrate at 213,000× *g* for 2 h at 4 °C using a Type 45 Ti rotor.
4. Resuspend the resulting OMV pellet in sterile PBS.
5. Prepare discontinuous OptiPrep gradient (40%, 35%, 30%, 25%, 20%) in Tricine-NaOH buffer (10 mM, pH 7.4, 0.85% sodium chloride).
6. Load the 40% OptiPrep gradient buffer with OMV suspension at bottom of tube and sequentially layer other concentrations on top.
7. Ultracentrifuge at 135,000× *g* for 16 h at 4 °C using a SW 41Ti rotor.
8. Collect 1-mL fractions from top to bottom, and analyze fractions by SDS-PAGE to identify OMV-rich fractions.
9. Pool OMV-containing fractions and dilute at least tenfold in PBS (*see* **Note 2**).
10. Ultracentrifuge again at 200,500× *g* for 2 h at 4 °C.
11. Resuspend the purified OMVs in sterile PBS. Store at −80 °C for further use (*see* **Note 3**).

3.3 Transmission Electron Microscopy (TEM)

1. Negative stain with 1.5% aqueous uranyl acetate on 300 mesh carbon-coated copper grids.
2. Acquire image via Thermo-Fisher Tecnai Spirit Twin STEM equipped with a SIS MegaView III camera in TEM mode at 120 kV.

3.4 Dynamic Light Scattering (DLS)

1. Dilute OMVs to 100 ng/μL in PBS.
2. Analyze with Malvern Zetasizer Nano ZS and record Z-average sizes and their distribution.

3.5 SDS-PAGE and In-Gel Digestion

1. Load 5 ~ 10 μg OMVs onto 12% SDS-PAGE. Run electrophoresis under standard conditions.
2. Stain gel with Coomassie Brilliant Blue, followed by destaining using the destaining solution.
3. Excise gel into ~2–4 gel bands per sample, and cut into approximately 1 mm^3 cubes using a clean scalpel.
4. Wash gel pieces sequentially with 200-μL ultrapure water, and then dehydrate in 1:1 (v/v) 100-mM ammonium bicarbonate and acetonitrile, followed by 100% acetonitrile until white.
5. Add 70 μL of reduction buffer and incubate 1 h at 56 °C.
6. Add 100 μL of alkylation buffer, and incubate at room temperature for 1 h in the dark.
7. Rewash the gel slices with 100-mM ammonium bicarbonate, and acetonitrile (1:1 v/v). Dry gel slices completely in a SpeedVac or a vacuum concentrator.
8. Rehydrate dry gel pieces with 50 μL of trypsin digestion buffer and incubate at 37 °C overnight.
9. Extract tryptic peptides with 70 μL of peptide extraction solution 1 and once with 70-μL peptide extraction solution 2. Combine all extracts from each sample.
10. Lyophilize the peptide extract, resuspend in 20 μL of final resuspension solution, and store at −20 °C if not analyzed immediately.

3.6 LC-MS/MS Analysis

1. Carry out LC-MS/MS analysis on an orbitrap fusion tribrid mass spectrometer (Thermo Fisher Scientific) equipped with a nanospray Flex ion source and coupled to a Dionex UltiMate 3000 RSL Cnano system.
2. Load peptides (5–15 μL) onto a PepMap C18 trap column (5 μm, 100 μm i.d. × 20 mm) at a flow rate of 20 μL/min.
3. Perform chromatographic separation on an analytic PepMap C18 column (2 μm, 75 μm i.d. × 25 cm) at 35 °C using the following gradient: 0–120 min: linear increase from 5% to 38%

acetonitrile in 0.1% formic acid, followed by ramping to 90% and washing.

Maintain flow rate of 300 nL/min.

4. Operate the mass spectrometer in the positive ion mode, with a spray voltage of 1.6 kV and a capillary temperature of 275 °C.
5. Acquire MS1 survey scans in the Orbitrap at a resolution of 120,000, mass range m/z 375–1575.
6. Acquire data-dependent MS2 scans in the ion trap (top speed 3-s cycle).
7. Perform fragmentation with collision-induced dissociation at 30% normalized collision energy, isolation window of 1.6 m/z, and dynamic exclusion of 40 s with a ± 10 ppm exclusion window.
8. Acquire data using Xcalibur software (Thermo Fisher Scientific).

3.7 Data Analysis and Protein Annotation

1. Analyze raw data using Proteome Discoverer 2.2 with the Sequest HT algorithm.
2. Perform a database search against the *F. nucleatum* RefSeq protein database (2416 entries from UniProt).
3. Search parameters include the following:
   (a) Enzyme: trypsin (maximum two missed cleavages).
   (b) Precursor mass tolerance: 10 ppm.
   (c) Fragment ion mass tolerance: 0.6 Da.
   (d) Fixed modification: carbamidomethylation (C).
   (e) Variable modifications: oxidation (M), deamidation (N/Q), and N-terminal acetylation.
4. Filter high-confidence peptide identifications at 1% FDR using Percolator.
5. Calculate protein abundance based on precursor ion intensities, normalized by molecular weight. Relative molar abundance (%) is calculated as follows:

   (Abundance of individual protein / Total protein abundance) × 100%
6. Perform functional and localization annotations using the following:
   (a) Vaxign and CELLO2go for subcellular localization and adhesion prediction.
   (b) PRED-TMBB2 for β-barrel outer membrane prediction.
   (c) LipoP 1.0 for signal peptide and lipoprotein classification.
   (d) InterProScan for protein domain analysis.
   (e) SecretomeP 2.0 for nonclassical secretion pathway prediction.

7. Conduct antigenic region prediction using the following:
   (a) EMBOSS Antigenic for linear epitope mapping.
   (b) BCPREDS for B-cell epitope prediction.
   (c) Vaxitope for T-cell epitope prediction.
8. Perform pathway and functional classification using the following:
   (a) KEGG for metabolic pathway mapping.
   (b) EggNOG 4.5.1 for orthologous group and functional category assignment.
9. Carry out visualization and comparative analysis in R using ggplot2.

4 Notes

1. Anaerobic culture conditions: *F. nucleatum* should be cultured strictly under anaerobic conditions. Use an anerobic chamber of a gas-generating sachet system. To ensure a fully anaerobic environment, culture media should be placed inside the anaerobic chamber at least 1 day in advance to eliminate residual oxygen.
2. Fraction collection: After gradient centrifugation, carefully collect 1 mL of fractions from top to bottom without disturbing the interface. OMVs are usually enriched in the 25–30% fraction (need further verification by SDS-PAGE or protein concentration).
3. Store OMV aliquots at −80 °C to prevent degradation caused by repeated freeze-thaw cycles. For optimal preservation of vesicles integrity, flash-freeze aliquots in liquid nitrogen before storage.

Conflict of Interest

The authors have no conflicts of interest to declare that are relevant to the content of this chapter.

References

1. Schwechheimer C, Kuehn MJ (2015) Outer-membrane vesicles from Gram-negative bacteria: biogenesis and functions. Nat Rev Microbiol 13(10):605–619. https://doi.org/10.1038/nrmicro3525
2. O'Ryan M, Stoddard J, Toneatto D, Wassil J, Dull PM (2014) A multi-component meningococcal serogroup B vaccine (4CMenB): the clinical development program. Drugs 74(1):15–30. https://doi.org/10.1007/s40265-013-0155-7
3. Bachmann MF, Jennings GT (2010) Vaccine delivery: a matter of size, geometry, kinetics and molecular patterns. Nat Rev Immunol

10(11):787–796. https://doi.org/10.1038/nri2868

4. Bolstad AI, Jensen HB, Bakken V (1996) Taxonomy, biology, and periodontal aspects of *Fusobacterium nucleatum*. Clin Microbiol Rev 9(1):55–71. https://doi.org/10.1128/CMR.9.1.55
5. Isomura K (1985) Bacterial endotoxin from *Fusobacterium nucleatum*--particularly the role of endotoxin in periodontal diseases. Nihon Shishubyo Gakkai Kaishi 27(2): 328–351
6. Strauss J, White A, Ambrose C, McDonald J, Allen-Vercoe E (2008) Phenotypic and genotypic analyses of clinical *Fusobacterium nucleatum* and *Fusobacterium periodonticum* isolates from the human gut. Anaerobe 14(6): 301–309. https://doi.org/10.1016/j.anaerobe.2008.12.003
7. Yang NY, Zhang Q, Li JL, Yang SH, Shi Q (2014) Progression of periodontal inflammation in adolescents is associated with increased number of *Porphyromonas gingivalis*, *Prevotella intermedia*, *Tannerella forsythensis*, and *Fusobacterium nucleatum*. Int J Paediatr Dent 24(3):226–233. https://doi.org/10.1111/ipd.12065
8. Strauss J, Kaplan GG, Beck PL, Rioux K, Panaccione R, Devinney R, Lynch T, Allen-Vercoe E (2011) Invasive potential of gut mucosa-derived *Fusobacterium nucleatum* positively correlates with IBD status of the host. Inflamm Bowel Dis 17(9):1971–1978. https://doi.org/10.1002/ibd.21606
9. Flanagan L, Schmid J, Ebert M, Soucek P, Kunicka T, Liska V, Bruha J, Neary P, Dezeeuw N, Tommasino M, Jenab M, Prehn JH, Hughes DJ (2014) *Fusobacterium nucleatum* associates with stages of colorectal neoplasia development, colorectal cancer and disease outcome. Eur J Clin Microbiol Infect Dis 33(8):1381–1390. https://doi.org/10.1007/s10096-014-2081-3
10. Liu J, Hsieh CL, Gelincik O, Devolder B, Sei S, Zhang S, Lipkin SM, Chang YF (2019) Proteomic characterization of outer membrane vesicles from gut mucosa-derived *Fusobacterium nucleatum*. J Proteome 195:125–137. https://doi.org/10.1016/j.jprot.2018.12.029

Chapter 13

An Asymmetric Gas Coculture System for Oral Pathobionts and Gingival Epithelial Cells

Zeyang Pang, Collins I. Chimezie, Amber E. Haley, Nicole M. Cady, Lujia Cen, Thomas M. Schmidt, Xuesong He, and Jiahe Li

Abstract

Establishing a coculture system that models oral microbial-host interaction can be challenging. The oral cavity harbors a mixture of both anaerobic and aerobic bacteria and is subjected to alternating periods of rich oxygen exposure and oxygen-limited conditions. Here, we model an asymmetric coculture system comprising two oral anaerobic pathobionts—*Fusobacterium nucleatum* and *Porphyromonas gingivalis*. This system not only preserves the physiological conditions of both the bacteria and oral gingival epithelial cells but also effectively simulates the behavioral dynamics of pathogen-epithelial cell interactions under physiologically relevant oxygen levels.

Key words Oral anaerobes, Asymmetric coculture, Oral epithelial cells, *Fusobacterium nucleatum*, *Porphyromonas gingivalis*, Host-microbe interactions

1 Introduction

Establishing a suitable environment for the growth of evolutionarily distinct cells is essential for gaining a comprehensive understanding of physiological interactions between host tissues and microbes [1]. The mammalian oral cavity represents a particularly complex ecosystem, characterized by dynamic fluctuations in oxygen availability and continuous interactions among commensal, opportunistic, and pathogenic bacteria [2]. While mouse models fail to capture key differences in oral microbial structure and ecology between mice and humans, several advanced in vitro systems have been developed to recapitulate aspects of human host-microbe interactions in a reductionist yet physiologically relevant manner. For example, microfluidic and chip-based platforms, such as the gingival crevice-on-chip, have enabled modeling of crevicular fluid flow and host-oral microbiome interactions [3], while three-dimensional humanized gingival pocket models have provided

Hung Ton-That (ed.), *Fusobacterium nucleatum: Methods and Protocols*, Methods in Molecular Biology, vol. 3055, https://doi.org/10.1007/978-1-0716-5388-3_13,

insight into oral microbial colonization and host tissue responses in a more physiologically structured setting [4]. Complementing these complex systems, simpler coculture platforms have also been employed to interrogate pathogen-host interactions under anaerobic conditions. For instance, *P. gingivalis* has been cocultured with human umbilical vein endothelial cells (HUVECs) in an anaerobic chamber, directly linking periodontal pathogens with vascular cell responses [5]. However, it remains questionable whether maintaining host cells under anaerobic conditions is physiologically relevant, despite no obvious loss of cell viability. Collectively, these approaches underscore the need for versatile and physiologically relevant coculture models that can accommodate both anaerobic oral bacteria and oxygen-dependent host cells.

To address this need and better depict how microbiota-host interactions contribute to disease, we recently developed a coculture system that overcomes limitations of previous strategies by maintaining distinct yet interacting oxygen environments [6]. This system enables long-term coculture of oral anaerobes such as *F. nucleatum* and *P. gingivalis* with oral epithelial cells under conditions that closely mimic the natural oral cavity. Herein, we present a detailed protocol describing this system.

2 Materials

2.1 Preparation of Gingival Keratinocyte 2D Monolayers

1. Transwell inserts with the cell growth area of 0.33 cm^2, equipped with a transparent polyethylene terephthalate membrane measuring 6.5 mm in diameter and featuring 0.4 μm pores (e.g., VWR 76313-906).
2. An hTERT-immortalized gingival keratinocyte (TIGK) cell line (e.g., ATCC CRL-3397).
3. 100 mM of acetic acid: 30 μL of glacial acetic acid in 5 mL of sterile water; sterile-filtered.
4. Collagen IV stock: 1 mg/mL of human placenta, Bornstein and Traub type IV, in 100 mM of acetic acid (*see* **Note 1**).
5. 37 °C cell culture incubator.
6. Keratinocyte Growth Kit: 0.4% extract P, 0.5 ng/mL of rh TGF-alpha, 6 mM of L-glutamine, 100 ng/mL of hydrocortisone, 5 μg/mL of insulin, 1.0 μM of epinephrine, and 5 μg/mL of apo-transferrin.
7. Dermal cell basal medium (DCBM).
8. Dulbecco's modified Eagle's medium (DMEM).
9. Fetal bovine serum (FBS).
10. 100 U/mL of penicillin-streptomycin.
11. Milli-Q water.
12. 0.22-μm syringe filters.

2.2 Asymmetric Coculture Setup

1. *F. nucleatum* strain (e.g., ATCC 23726 and ATCC 25586).
2. *P. gingivalis* strain (e.g., ATCC 33277).
3. Columbia broth (CB).
4. CB blood agar (5% sheep blood): dissolve 8.75 g of CB agar powder and 3.75 g of agar in distilled water, and adjust the final volume to 250 mL. Autoclave the mixture at 121 °C for 15 min, cool to approximately 50 °C, and then supplement with 12.5 mL of defibrinated sheep blood (*see* **Note 2**).
5. Anaerobic chamber (see Fig. 1).
6. Magnetic stirrer.
7. Coculture medium (CCM).

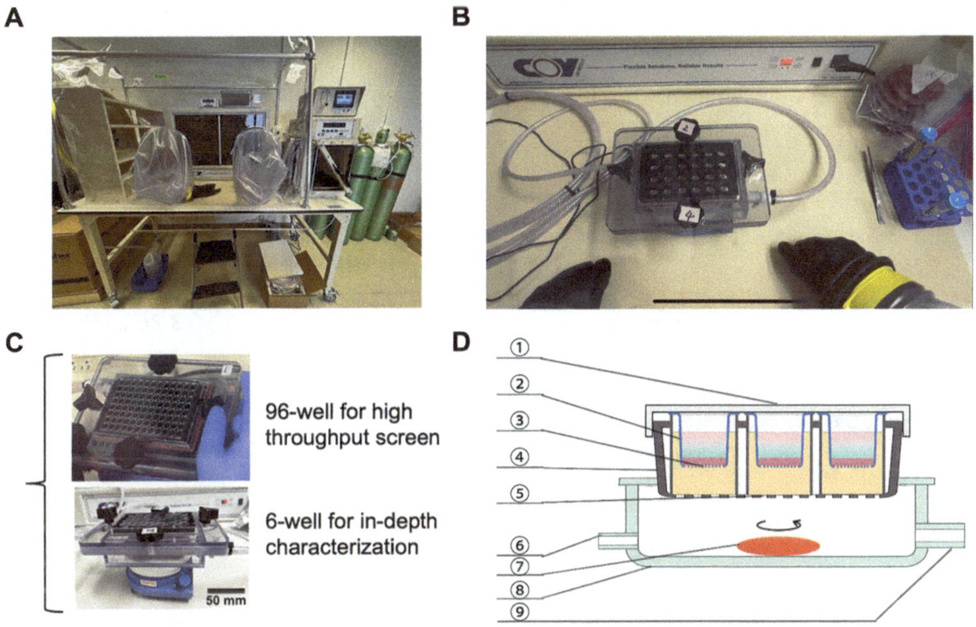

Fig. 1 Schematic representation of the asymmetric coculture system. (**a**) Photograph of the anaerobic chamber used to maintain asymmetric coculture conditions. (**b**) Top view of the asymmetric coculture chamber, showing the gas-permeable plate and the gas inlet and outlet ports. (**c**) The gas-permeable plate can be substituted with a standard 96-well or 24-well plate, depending on experimental needs. (**d**) The gas-permeable plate (④) is enclosed within a protective plate cap (①) to reduce contamination risk and prevent gas leakage. Transwell inserts (②) are positioned on the plate, with their apical chambers containing coculture medium (CCM) to support both an epithelial cell monolayer (③) and oral anaerobic bacteria. The basolateral compartment contains conventional cell culture medium. A gas-permeable membrane (⑤) at the base of the plate enables unidirectional oxygen diffusion from the basolateral side. Basolateral gas, adjusted to 10% O_2, enters through the inlet (⑥) and is evenly circulated throughout the chamber (⑧) by a magnetic stirrer (⑦). Exhaust gas exits through the outlet (⑨), maintaining a continuous and controlled oxygen gradient. (Illustration created with BioRender. *Figure adapted from Pang* et al., *International Journal of Oral Science, 2025, licensed under CC BY 4.0*)

Reagents	Working concentrations
Mineral solutions	
Dipotassium phosphate (K_2HPO4)	0.51 mM
Potassium diphosphate (KH_2PO4)	2.94 mM
Potassium chloride (KCl)	5.37 mM
Magnesium sulfate heptahydrate ($MgSO_4 \cdot 7H_2O$)	0.41 mM
Calcium chloride dihydrate ($CaCl_2 \cdot 2H_2O$)	1.26 mM
Magnesium dichloride hexahydrate ($MgCl_2 \cdot 6H_2O$)	0.49 mM
Ammonium sulfate ($(NH_4)_2SO_4$)	3.4 mM
Sodium chloride	8 g/L
Vitamin mix	
Biotin	8.19 μM
Folic acid	4.53 μM
Pyridoxine HCl	48.63 μM
Thiamine HCl	14.82 μM
Riboflavin	13.26 μM
Nicotinic acid	40.61 μM
D-Pantothenic acid hemicalcium salt	20.98 μM
Vitamin B12	7.37 μM
p-Aminobenzoic acid (PABA)	36.46 μM
α-Lipoic acid (thioctic acid)	24.23 μM
Trace metals	
EDTA	17.1 μM
$FeSO_4 \cdot 7H_2O$	3.60 μM
$ZnSO_4 \cdot 7H_2O$	6.26 μM
$CuSO_4 \cdot 5H_2O$	350 nM
$CoCl_2 \cdot 6H_2O$	7.57 μM
$MnSO_4 \cdot H_2O$	29.58 μM
$NiCl_2 \cdot 6H_2O$	2.95 μM

(continued)

Reagents	Working concentrations
Glucose	0.9 g/L
Yeast extract	1.25 g/L
Tryptone	5 g/L
Hemin	2 g/L
Bromocresol purple	1 g/L
Sodium bicarbonate	0.338 g/L

2.3 Analysis of Bacterial Viability

1. 96-well plates.
2. Multichannel pipette.
3. Phosphate-buffered saline (PBS).
4. CB blood agar plates.
5. Anaerobic chamber.
6. Ziplock bags.

3 Methods

3.1 Preparation of Gingival Keratinocyte 2D Monolayers

1. Prepare the Transwell collagen coating solution by diluting the collagen IV stock to 33 μg/mL using ice-cold ultrapure water.
2. Dispense 100 μL of the diluted solution into each apical chamber of the Transwell inserts.
3. Incubate the entire plate at 4 °C for 30 min, and then transfer it to a 37 °C cell culture incubator for 4 h to allow collagen fibrillogenesis (*see* **Notes 3–6**). The coated plates can be stored for later use (*see* **Note 7**).
4. To generate 2D monolayers of gingival keratinocytes, seed each Transwell insert with approximately 200,000 cells in 100 μL DCBM supplemented with the components of the Keratinocyte Growth Kit (*see* **Note 8**).
5. Simultaneously add 600 μL of DCBM to the basolateral chamber of the Transwell and incubate for more than 20 h at a 37 °C in a cell culture incubator.
6. The following day, replace the culture medium with fresh DMEM supplemented with 10% FBS and 100 U/mL of penicillin-streptomycin to initiate differentiation. Continue differentiation for 5 days (*see* **Note 9** and Fig. 2 as an example).

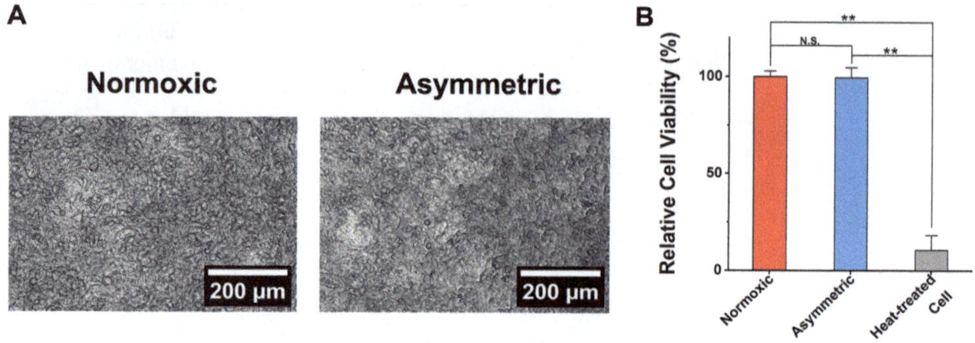

Fig. 2 Assessment of TIGK cell viability under coculture conditions. (**a**) Representative bright-field images of TIGK monolayers cultured in Transwell inserts under normoxic conditions or within the asymmetric coculture chamber for 24 h. Collagen coating reduces image clarity, as collagen fibers partially obscure fine cellular details compared with uncoated surfaces. (**b**) As an example, TIGK monolayer viability was preserved under both normoxic and asymmetric coculture conditions. Heat-treated cells were included as a negative control. Data are presented as mean ± SD ($n = 3$, $N = 3$). N.S., not significant ($p > 0.05$); **, $p < 0.01$. (*Figure adapted from Pang* et al., *International Journal of Oral Science, 2025, licensed under CC BY 4.0)*

3.2 Asymmetrical Coculture Setup

A recorded video demonstrating the setup is available here: https://drive.google.com/file/d/1daMtu_suAnRW4IsfOHozOqO9vjN_6uA0/view?usp=sharing.

1. Inoculate *F. nucleatum* (*Fn*) or *P. gingivalis* (*Pg*) from −80 °C frozen stocks into pre-reduced CB medium and culture overnight at 37 °C under anaerobic conditions (*see* Fig. 1a–c).
2. The following day, dilute *Fn* 1:100 and *Pg* 1:5 into 1 mL of CCM and continue incubation until the culture reaches an OD_{600} of ~0.4 (*see* **Note 10**).
3. On the day of the experiment, adjust and stabilize the anaerobic chamber temperature to 37 °C. Regenerate the catalyst at 120 °C for 2 h to ensure optimal anaerobic performance.
4. Transfer the gas-permeable culture plate into the anaerobic chamber via the airlock, and secure it within the asymmetrical coculture chamber using fixing screws.
5. Connect the gas inlet and outlet tubing to the chamber, and activate the magnetic stirrer. Supply 10% O_2-containing gas to the basolateral compartment at 5 standard cubic feet per hour (SCFH) for 10 s, then reduce to 0.5 SCFH (*see* **Note 11** and Fig. 1d).
6. Wash Transwell inserts three times with PBS, and then replace the medium with DMEM supplemented with 10% FBS and no antibiotics. Seal the 24-plate in a Ziplock bag, and then transfer the washed inserts into the asymmetric gas system through the airlock.

7. After a 3-h purge inside the anaerobic chamber (*see* **Note 12**), initiate coculture by aspirating the apical medium and replacing it with CCM or with bacterial suspensions (*Fn* or *Pg*) at multiplicity of infection (MOI) of 1 prepared in CCM (*see* **Note 13**).
8. Place the gas-permeable plate with Transwell inserts on the magnetic stirrer, and activate the system (*see* **Note 14**). Maintain the setup at 37 °C for 24 h for *Fn* or 48 h for *Pg*.

3.3 Analysis of Bacterial Viability

1. After the coculture is completed, prepare a serial dilution of the apical chamber solution by adding 90 μL of PBS to each well of a 96-well plate. Transfer 10 μL of the apical chamber solution into a well in Row A. Mix thoroughly, and then perform a tenfold serial dilution by transferring 10 μL from Row A to Row B. Repeat this process across subsequent rows until the desired dilution range is achieved.
2. Using a multichannel pipette, spot 2 μL of the bacterial suspensions from each row of the 96-well plate onto pre-reduced CB blood agar plates (*see* **Note 15**, Fig. 3, and Fig. 4 as examples). Allow the spots to fully air-dry.
3. Incubate the inverted CB blood agar plates at 37 °C inside the anaerobic chamber, placing them in a sealed Ziplock bag (*see* **Note 16**) with a wet paper towel to maintain humidity. Incubate for 2–3 days until colonies become visible.
4. Determine colony forming unit (CFU) counts by back-calculation based on the dilution factor and spotted volume.

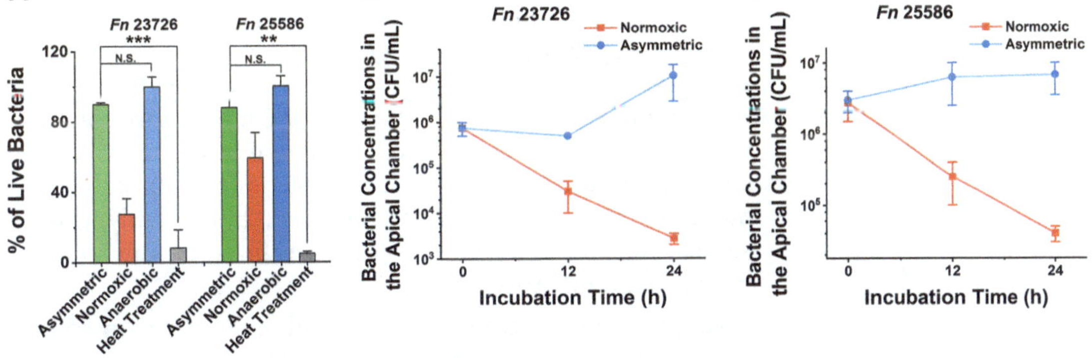

Fig. 3 Assessment of *F. nucleatum* viability and growth under asymmetric coculture conditions. (**a**) As an example, the viability of *F. nucleatum* strains 23726 and 25586 was quantified by flow cytometry following exposure to different treatment groups: asymmetric gas coculture, normoxic coculture, anaerobic culture (positive control), and heat-treated dead cells (negative control). Statistical significance was determined as follows: N.S., $p > 0.05$; **, $p < 0.01$; ***, $p < 0.001$ ($n = 2$ technical replicates, $N = 3$ biological replicates). CFUs of (**b**) *Fn* 23726 and (**c**) *Fn* 25586 were enumerated in the apical chamber under normoxic and asymmetric coculture conditions at 0, 12, and 24 h ($n = 2$, $N = 3$). The multiplicity of infection (MOI) was 1 for all assays. (*Figure adapted from Pang* et al., *International Journal of Oral Science, 2025, licensed under CC BY 4.0)*

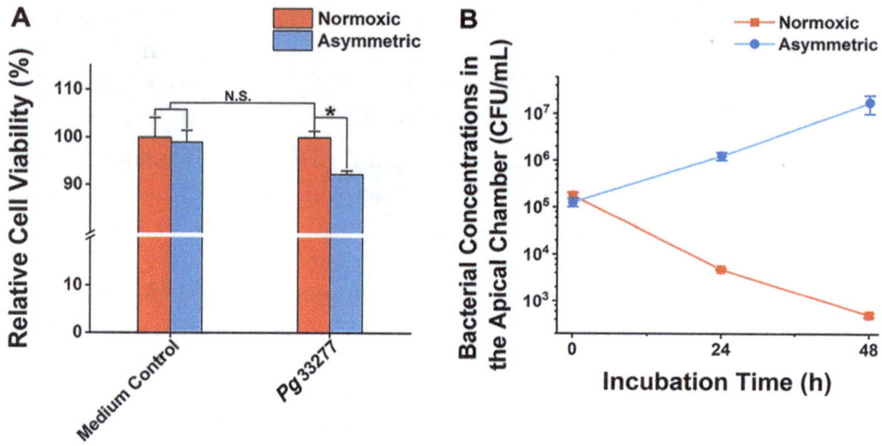

Fig. 4 Growth of *P. gingivalis* and TIGK cell viability under asymmetric coculture conditions. (**a**) Quantification of TIGK cell viability following asymmetric coculture with *P. gingivalis* 33277, based on the flow cytometry analysis shown above. *$p = 0.01–0.05$; N.S., not significant ($p > 0.05$). Biological replicates: $N = 3$; technical replicates: $n = 2$. (**b**) CFUs of *P. gingivalis* 33277 in the apical chamber under normoxic and asymmetric coculture conditions at 0, 24, and 48 h (MOI = 1; $n = 2$ technical replicates, $N = 3$ biological replicates). (*Figure adapted from Pang* et al., *International Journal of Oral Science, 2025, licensed under CC BY 4.0)*

4 Notes

1. Collagen IV (e.g., Sigma-Aldrich, C5533) is supplied in 5-mg vials, and preparation of the working solution requires careful handling. Retrieve a vial from the freezer, briefly equilibrate to room temperature, and then transfer it to ice in a sterile biosafety cabinet. Remove the outer metal seal and pierce the rubber stopper with a 28-gauge needle to release the internal vacuum. Add 5 mL of ice-cold 100-mM acetic acid to the vial. Gently rotate the vial at 4 °C for at least 5 h to ensure complete solubilization. Once dissolved, gently mix the collagen IV solution, and aliquot it into pre-cooled sterile tubes on ice. Store the aliquots at −80 °C until use. Importantly, do not sterile-filter the collagen IV solution, as filtration can compromise protein integrity.
2. Pour plates in a biosafety cabinet and allow them to solidify overnight. Prepared plates may be stored at 4 °C for up to 1 week before being transferred into the anaerobic chamber. Before being used in experiments, plates must be pre-reduced overnight under anaerobic conditions.
3. The collagen VI solution should be thawed at 4 °C to preserve its native triple-helical conformation and higher-order assemblies, which are highly sensitive to temperature fluctuations [7]. Thawing at low temperature minimizes disruption of the

hydrogen bonding and hydrophobic interactions that stabilize the protein, thereby preventing partial denaturation, premature fibril formation, and irreversible aggregation. Thawing above 4 °C can create localized thermal and pH gradients as ice melts, leading to nonuniform solubility, precipitation, or gelation, all of which compromise bioactivity. Maintaining the solution at ≤4 °C during thawing also suppresses residual protease activity and microbial growth, helping keep the structural and functional integrity of collagen VI for reproducible experimental use. Once fully thawed, the collagen solution may be stored at 4 °C for up to 2 weeks.

4. All collagen-coating steps should be performed on ice to maintain the collagen solution in its soluble state and prevent premature fibrillogenesis. Collagen molecules begin to self-assemble into fibrils at physiological temperatures, and even brief warming can initiate partial gelation, resulting in uneven coating and altered matrix architecture [8]. Working on ice slows molecular motion, stabilizes the triple-helical structure, and preserves the uniformity of the coating solution during handling. This temperature control ensures that collagen is applied evenly across the substrate surface, yielding a consistent and reproducible extracellular matrix layer that supports optimal cell adhesion and growth.
5. Vigorous pipetting or forceful agitation of the collagen solution should be avoided, as shear stress can disrupt its triple-helical structure and partially denature the protein, diminishing its ability to form a stable and biologically relevant fibrillar network. Excessive aeration during aggressive handling can also introduce bubbles, which interfere with uniform coating or gel formation and create regions of inconsistent thickness. Moreover, high shear forces may expose hydrophobic domains and promote localized aggregation, leading to uneven fibrillogenesis. Gentle handling preserves collagen's native conformation, solubility, and uniform distribution, ensuring reproducible structural and functional properties in downstream applications.
6. After adding the collagen solution to the apical chamber of the Transwell, keep the plate at 4 °C for 30 min before transferring it to a 37 °C incubator. This cooling period preserves the collagen in a fully liquid state during the critical initial settling phase, ensuring uniform spread and consistent coating thickness across the membrane. Premature warming can trigger partial fibrillogenesis or gelation before the solution distributes evenly, leading to an irregular matrix. Maintaining the solution at 4 °C promotes even coverage, while the subsequent shift to 37 °C initiates controlled fibril formation, producing a reproducible collagen layer suitable for cell attachment and culture.

7. After coating, the plates can be used immediately for cell seeding. For storage, add 500 μL of DPBS to a blank well adjacent to each unused Transwell to minimize evaporation. Seal the gaps in the plate containing unused Transwells with parafilm. When stored under these conditions, coated plates remain stable for up to 2 months.
8. After adding the cell suspension to the apical chamber of the Transwell, avoid tilting, tapping, or moving the plate. Instead, transfer it directly and steadily to the incubator. Any movement during this early phase can disturb the settling cells, creating uneven distribution and localized variations in cell density. Such inconsistencies can compromise monolayer formation, resulting in variable barrier properties and reduced reproducibility in downstream assays.
9. Supplementing DMEM with Ca^{2+} promotes TIGK differentiation by mimicking the physiological extracellular calcium environment required for epithelial maturation [9]. Elevated Ca^{2+} enhances cadherin-dependent cell-cell adhesion by stabilizing desmosomes and adherens junctions, supporting the development of a stratified epithelial architecture [10]. In TIGKs, this calcium-mediated signaling drives the transition from a proliferative, basal-like phenotype toward a more differentiated state, closely reflecting the in vivo oral epithelial barrier. Thus, physiological calcium levels provide both the biochemical cues and adhesion stability necessary for controlled epithelial differentiation.
10. For more oxygen-tolerant species such as *F. nucleatum*, an overnight culture is typically sufficient to reach and OD_{600} of 0.4. In contrast, oxygen-sensitive anaerobes such as *P. gingivalis* grow more slowly under strictly anaerobic conditions and may require 36 h or longer to reach the same optical density.
11. A brief high-flow gas flush is used at the start of the procedure to rapidly displace residual oxygen from the coculture chamber. After this initial purge, the flow rate is reduced to a controlled, moderate level that ensures adequate oxygen delivery to the basolateral cell monolayer while maintaining low oxygen concentrations in the apical compartment, thereby preventing interference with anaerobic bacterial growth.
12. The Transwell inserts containing the cell monolayer are exposed to oxygen-containing carrier gas for an extended equilibration period. This gradual aeration allows the monolayer's microenvironment to transition smoothly toward the composition of the carrier gas, minimizing physiological stress and preserving cellular viability.

13. At an MOI of 1, the bacterial load generated over 48 h of anaerobic culture remains within the medium's buffering capacity and does not cause significant pH deviations.
14. To reduce hypoxia resulting from limited oxygen diffusion and cellular oxygen consumption, a magnetic stirrer is employed to enhance gas-liquid exchange. In static cultures, diffusion across the liquid-air interface is often insufficient to meet the metabolic demands of confluent epithelial monolayers, leading to localized oxygen depletion. Gentle stirring promotes mixing, decreases the thickness of the unstirred boundary layer, and accelerates the replenishment of dissolved oxygen from the gas phase. This results in more uniform oxygenation and minimizes artifacts associated with diffusion-limited microenvironments.
15. CB blood agar plates are preincubated in the anaerobic chamber for at least 12 h to remove residual oxygen. After equilibration, the plates exhibit a characteristic dark-red coloration due to hemoglobin reduction. Because CB blood agar contains relatively high moisture content, plates are subsequently air-dried until the surface appears dry to prevent colony spreading and ensure optimal bacterial growth.
16. Blood agar plates are best stored in the dark because light exposure accelerates the oxidation and degradation of hemoglobin and other blood components. Prolonged illumination can lead to discoloration of the medium, reduced stability of red blood cells, altered background appearance, and interference with the interpretation of hemolysis. Avoiding light helps maintain the medium's integrity, appearance, and diagnostic reliability.

Acknowledgments

This work was supported by National Science Foundation CAREER (2238972) and National Institute of Dental and Craniofacial Research awards (R03DE031329 and R01DE030943). We thank the Translational Tissue Modeling Laboratory and the Microscopy Core at the University of Michigan, Ann Arbor, for their valuable support.

Conflict of Interest The authors have no conflicts of interest to declare that are relevant to the content of this chapter.

References

1. Baker JL, Mark Welch JL, Kauffman K et al (2024) The oral microbiome: diversity, biogeography and human health. Nat Rev Microbiol 22(2):89–104. https://doi.org/10.1038/s41579-023-00963-6
2. Welch JLM, Rossetti BJ, Rieken CW et al (2016) Biogeography of a human oral microbiome at the micron scale. Proc Natl Acad Sci USA 113(6):E791–E800. https://doi.org/10.1073/pnas.1522149113
3. Makkar H, Zhou Y, Tan KS et al (2023) Modeling crevicular fluid flow and host-oral microbiome interactions in a gingival crevice-on-chip. Adv Healthc Mater 12(6):1–13. https://doi.org/10.1002/adhm.202202376
4. Adelfio M, Martin-Moldes Z, Erndt-Marino J et al (2023) Three-dimensional humanized model of the periodontal gingival pocket to study oral microbiome. Adv Sci 10(12):1–13. https://doi.org/10.1002/advs.202205473
5. Wunsch CM, Lewis JP (2015) *Porphyromonas gingivalis* as a model organism for assessing interaction of anaerobic bacteria with host cells. J Vis Exp 2015(106):1–12. https://doi.org/10.3791/53408
6. Pang Z, Cady NM, Cen L et al (2025) Physiologically relevant coculture model for oral microbial-host interactions. Int J Oral Sci 17(1):1–12. https://doi.org/10.1038/s41368-025-00365-9
7. Cescon M, Gattazzo F, Chen P et al (2015) Collagen VI at a glance. J Cell Sci 128(19):3525–3531. https://doi.org/10.1242/jcs.169748
8. Parenteau-Bareil R, Gauvin R, Berthod F (2010) Collagen-based biomaterials for tissue engineering applications. Materials 3(3):1863–1887. https://doi.org/10.3390/ma3031863
9. McClintock SD, Attili D, Dame MK et al (2020) Differentiation of human colon tissue in culture: effects of calcium on trans-epithelial electrical resistance and tissue cohesive properties. PLoS One 15(3):1–18. https://doi.org/10.1371/journal.pone.0222058
10. Chen T, Karedla N, Enderlein J (2024) Observation of E-cadherin adherens junction dynamics with metal-induced energy transfer imaging and spectroscopy. Commun Biol 7(1):1–10. https://doi.org/10.1038/s42003-024-07281-4

Chapter 14

A 3D Tumor Spheroid Platform to Model *Fusobacterium nucleatum*-Induced Tumorigenesis

Yi-Wei Chen, Omri Kariv, and Hung Ton-That

Abstract

Cancer cell line spheroids provide a three-dimensional (3D) culture platform that enables the evaluation of cancer cell aggregation, tumor initiation, cell proliferation, and overall tumor growth. Compared with conventional two-dimensional monolayer cultures, spheroid models more faithfully recapitulate the architectural, physiological, and microenvironmental features of solid tumors, making them particularly well suited for studying tumor biology and microbe-host interactions. Previous work has leveraged this 3D system to investigate the role of the oral oncobacterium *Fusobacterium nucleatum* in promoting tumorigenesis. Here, we present a detailed protocol in which colorectal cancer cells are seeded into ultralow attachment plates and co-cultured with *F. nucleatum* either prior to spheroid formation, to model early tumor development, or after spheroids have been established, to model tumor progression. This assay provides a tractable platform for dissecting cancer cell-microbe interactions and supports a contributory role for *F. nucleatum* in colorectal tumor development.

Key words *Fusobacterium nucleatum*, Virulence, Spheroid formation, Spheroid growth, Colorectal cancer

1 Introduction

Fusobacterium nucleatum is an anaerobic, Gram-negative bacterium commonly residing in the human oral cavity, where it plays a central role in the development of periodontal diseases such as gingivitis and periodontitis [1–3]. Beyond its established role in oral pathology, *F. nucleatum* has been increasingly implicated in a variety of extra-oral conditions, including adverse pregnancy outcomes and multiple malignancies such as colorectal, breast, and pancreatic cancers [4–9]. Notably, several studies have reported an enrichment of *Fusobacterium* spp. in oral squamous cell carcinoma tissues compared with adjacent healthy tissues [10–12]. Recognized as both an opportunistic pathogen and an oncobacterium [5], *F. nucleatum* is capable of disseminating to distal body sites—including placental, breast, pancreatic, and colorectal

Hung Ton-That (ed.), *Fusobacterium nucleatum: Methods and Protocols*, Methods in Molecular Biology, vol. 3055, https://doi.org/10.1007/978-1-0716-5388-3_14,

tissues—where it contributes to processes such as preterm birth [4], tumor initiation, and tumor proliferation [6, 7, 9]. Importantly, the presence of *F. nucleatum* in colorectal cancer (CRC) tissues has been associated with poorer clinical outcomes and reduced overall survival [13, 14].

Although traditional two-dimensional (2D) cell culture models have enabled major advances in cancer research, they fall short in recapitulating the architectural complexity and microenvironmental features of solid tumors [15]. To better model host-microbe interactions, previous studies have employed three-dimensional (3D) spheroid systems to investigate interactions between *F. nucleatum*, host immune cells, and microbial factors influencing tumor biology [16, 17]. Here, we present a detailed protocol for generating CRC tumor spheroids using ultralow attachment plates to evaluates the effects of *F. nucleatum* on tumor formation and growth. This 3D spheroid platform provides a more physiologically relevant model for studying the mechanisms by which fusobacterial factors shape cancer development and progression.

2 Materials

All solutions should be prepared in sterile and molecular grade water.

2.1 Cell Culture for Spheroid

1. Cancer cell line HCT116.
2. 100-mm tissue culture plates.
3. 6-well ultralow attachment plate.
4. Gibco McCoy's 5A cell culture medium.
5. Gibco fetal bovine serum (FBS).
6. 5A+ (45 mL of Gibco McCoy's 5A cell culture medium mixed with 5 mL of Gibco FBS).
7. Phosphate buffered saline (PBS).
8. 0.25% trypsin.
9. Hemocytometer.
10. Optical microscope.
11. 10-mL serological pipettes, 1.5-mL microtubes, and 15-mL centrifuge tubes.
12. Benchtop centrifuge.
13. CO_2 cell incubator (5% CO_2 at 37 °C).

2.2 Preparation of F. nucleatum for Infection

1. *F. nucleatum* ATCC 23726 strain.
2. BD Bacto Tryptic Soy Broth.
3. BD Gibco Peptone.

4. TSP medium: Mixed 3 g of BD Bacto Tryptic Soy Broth and 1 g of BD Gibco Peptone in 100-mL water; autoclave the medium before use.
5. Cysteine.
6. Hungate tubes (16 × 125 mm).
7. Coy anaerobic chamber (95% N_2, 2.5% H_2, and 2.5% CO_2; 25 °C and 37 °C).
8. 10% cysteine solution: Dissolve 1 g of cysteine in 10-mL water, and store in a Hungate tube; degas the solution in an anaerobic chamber; seal the Hungate tube; autoclave the solution before use.
9. TSPC medium: Mix 20 mL of TSP medium with 0.1 mL of 10% cysteine.
10. Glass culture tubes (16 × 125 mm).
11. PBS.
12. Benchtop microtube centrifuge.
13. 15-mL serological pipettes and 1.5-mL microtubes.

2.3 Setting of Spheroid Formation and Growth Assays

1. CO_2 cell incubator (5% CO_2; 37 °C).
2. Optical microscope and CCD.
3. Software for spheroid measurement such as ImageJ.

3 Methods

3.1 Cell Culture for Spheroid Formation

1. Seed 1×10^6 HCT116 cells to a 100-mm tissue culture plate in 10 mL of 5A+ medium, and incubate the cells in a 37 °C, 5% CO_2 incubator.
2. Once the culture reaches around 85% confluency, aspirate the medium and wash the cells twice with PBS.
3. Remove PBS and add 1 mL of 0.25% trypsin to detach the cells (*see* **Note 1**). Incubate the plate at 37 °C for 2 min to facilitate cell detachment (*see* **Note 2**).
4. Verify complete cell detachment under an optical microscope.
5. Neutralize the trypsin by adding 1 mL of 5A+ medium.
6. Transfer the cell suspension into a 15-mL centrifuge tube, and centrifuge under 1000× *g* for 3 min in a benchtop centrifuge.
7. Aspirate the supernatant and resuspend the cell pellet in 3 mL of 5A+ medium.
8. Pipette 10 μL of the cell suspension onto a hemocytometer, and perform cell counting using an optical microscope.
9. Seed 1×10^5 cells in 2 mL of 5A+ medium into each well of a 6-well ultralow attachment plate (*see* **Notes 3** and **4**).

3.2 Preparation of *F. nucleatum* for Infection

1. Inoculate *F. nucleatum* ATCC 23726 into 5 mL of TSPC medium, and incubate at 37 °C for 16–18 h in an anaerobic chamber maintained at 95% N_2, 2.5% H_2, and 2.5% CO_2 (*see* **Note 5**).
2. Transfer 1 mL of the overnight culture to a microcentrifuge tube, and centrifuge at 8000× *g* for 3 min.
3. Carefully aspirate the supernatant and wash the bacterial pellet twice with PBS.
4. Resuspend the washed pellet with PBS to prepare a bacterial suspension at the concentration of 5×10^8 CFU/mL.

3.3 Setup of the Spheroid Formation Assay

1. Add 10 μL of the bacterial suspension prepared in Subheading 3.2 to each cell-seeded well of the ultralow attachment plate from Subheading 3.1 to achieve an MOI of 50 (for the negative control without bacteria, add 10 μL of PBS instead) (*see* **Note 4**).
2. Gently mix the cells and bacteria by pipetting with a 1-mL pipette.
3. Place the plate in a 37 °C, 5% CO_2 incubator.
4. After 24 h, capture images of each well using an optical microscope equipped with a CCD camera (see examples in Fig. 1a).

3.4 Setup of the Spheroid Growth Assay

1. Allow spheroids to form without bacterial exposure by incubating the plate prepared in Subheading 3.1 at 37 °C in a CO_2 cell incubator for 1 day.

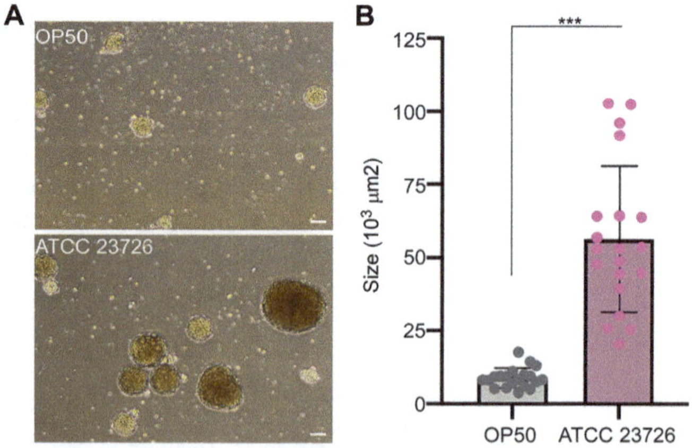

Fig. 1 *F. nucleatum*-induced cancer spheroid formation. A total of 1×10^5 cancer cells were seeded into a six-well ultralow attachment plate along with *F. nucleatum* at an MOI of 50. After 24 h, representative spheroid images were acquired (A), and spheroid size (A) was quantified using ImageJ. Statistical significance was determined using Student's t-test (***, $P < 0.001$). Scale bar: 100 μm

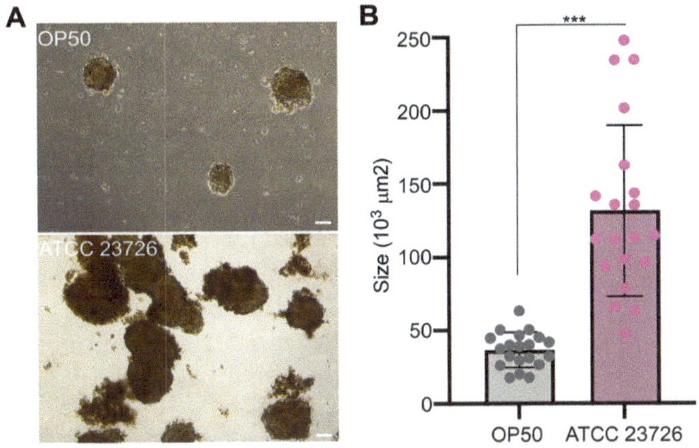

Fig. 2 *F. nucleatum*-induced cancer spheroid growth. A total of 1 × 10^5 cancer cells were seeded into a six-well ultralow attachment plate and allowed to form spheroids for 24 h before adding *F. nucleatum* at an MOI of 50. Following an additional 24 h of co-incubation, representative spheroid images were collected (A), and spheroid size (A) was quantified using ImageJ. Statistical significance was assessed using Student's t-test (***, $P < 0.001$). Scale bar: 100 µm

2. Add 10 µL of the bacterial suspension prepared in Subheading 3.2 to each well to achieve an MOI of 50 (for the negative control, add 10 µL of PBS instead) (*see* **Note 4**).
3. Gently mix the spheroid and bacteria by pipetting using 1-mL pipette.
4. Return the plate to a 37 °C, 5% CO_2 incubator.
5. After 24 h, acquire images of each well by an optical microscope equipped with a CCD camera (*see* Fig. 2a).
6. Import the spheroid images from both the spheroid formation and growth assays into ImageJ.
7. Measure the length (L) and width (W) of each spheroid in every well.
8. Calculate spheroid volume using the following equation: ½ × L × W × W (*see* Figs. 1b and 2b).

4 Notes

1. During trypsinization, 1 mL of 0.25% trypsin is sufficient to cover all cells in a 100-mm tissue culture plate. Ensure that the trypsin thoroughly contacts the monolayer by gently tilting the plate or lightly pipetting across the surface. Using excessive trypsin or prolonged exposure can damage the cells and compromise downstream assays.

2. The exposure time required for trypsinization is cell line-dependent. For HCT116 cells, approximately 2 min is typically adequate for complete detachment. Other lines, such as HT-29, may require up to 5 min. When working with a new cell line, monitor detachment frequently under a microscope to establish the optimal incubation time.
3. Experimental conditions may require optimization when using different cell lines or bacterial strains. This protocol provides a general framework for initiating spheroid assays with *F. nucleatum*. It is recommended to begin with the described conditions and adjust parameters as needed for specific applications.
4. When altering the MOI or the assay duration, it may be necessary to adjust both cell density and bacterial CFU. Beginning with a lower cell number (e.g., 1×10^5 cells) is recommended for consistent spheroid formation.
5. Prepare TSPC medium fresh immediately before use. Cysteine functions both as a nutrient for *Fusobacterium* and a reducing agent due to its free thiol group (–SH), helping maintain anaerobic conditions. Because cysteine oxidizes readily, add it to TSP medium just before inoculation to preserve reducing activity.

Acknowledgments

We thank our laboratory members for the critical review of this manuscript. Work reported in this publication was supported by the National Institute of Dental and Craniofacial Research (NIDCR) of the National Institutes of Health (NIH) under the award number DE033900 (to H.T.-T.).

Conflict of Interest The authors have no conflicts of interest to declare that are relevant to the content of this chapter.

References

1. Han YW (2015) *Fusobacterium nucleatum*: a commensal-turned pathogen. Curr Opin Microbiol 23:141–147. https://doi.org/10.1016/j.mib.2014.11.013
2. Danielsen B, Wilton JM, Baelum V, Johnson NW, Fejerskov O (1993) Serum immunoglobulin G antibodies to *Porphyromonas gingivalis*, *Prevotella intermedia*, *Fusobacterium nucleatum* and *Streptococcus sanguis* during experimental gingivitis in young adults. Oral Microbiol Immunol 8:154–160. https://doi.org/10.1111/j.1399-302x.1993.tb00658.x
3. Liu P, Liu Y, Wang J, Guo Y, Zhang Y, Xiao S (2014) Detection of *Fusobacterium nucleatum* and *fadA* adhesin gene in patients with orthodontic gingivitis and non-orthodontic periodontal inflammation. PLoS One 9:e85280. https://doi.org/10.1371/journal.pone.0085280

4. Han YW, Shen T, Chung P, Buhimschi IA, Buhimschi CS (2009) Uncultivated bacteria as etiologic agents of intra-amniotic inflammation leading to preterm birth. J Clin Microbiol 47:38–47. https://doi.org/10.1128/JCM.01206-08
5. Brennan CA, Garrett WS (2019) *Fusobacterium nucleatum* - symbiont, opportunist and oncobacterium. Nat Rev Microbiol 17:156–166. https://doi.org/10.1038/s41579-018-0129-6
6. Rubinstein MR, Wang X, Liu W, Hao Y, Cai G, Han YW (2013) *Fusobacterium nucleatum* promotes colorectal carcinogenesis by modulating E-cadherin/beta-catenin signaling via its FadA adhesin. Cell Host Microbe 14:195–206. S1931-3128(13)00260-6
7. Parhi L, Alon-Maimon T, Sol A, Nejman D, Shhadeh A, Fainsod-Levi T, Yajuk O, Isaacson B, Abed J, Maalouf N, Nissan A, Sandbank J, Yehuda-Shnaidman E, Ponath F, Vogel J, Mandelboim O, Granot Z, Straussman R, Bachrach G (2020) Breast cancer colonization by *Fusobacterium nucleatum* accelerates tumor growth and metastatic progression. Nat Commun 11:3259. https://doi.org/10.1038/s41467-020-16967-2
8. Gaba FI, Gonzalez RC, Martinez RG (2022) The role of oral *Fusobacterium nucleatum* in female breast cancer: a systematic review and meta-analysis. Int J Dent 2022:1876275. https://doi.org/10.1155/2022/1876275
9. Udayasuryan B, Ahmad RN, Nguyen TTD, Umana A, Monet Roberts L, Sobol P, Jones SD, Munson JM, Slade DJ, Verbridge SS (2022) *Fusobacterium nucleatum* induces proliferation and migration in pancreatic cancer cells through host autocrine and paracrine signaling. Sci Signal 15:eabn4948. https://doi.org/10.1126/scisignal.abn4948
10. Nagy KN, Sonkodi I, Szoke I, Nagy E, Newman HN (1998) The microflora associated with human oral carcinomas. Oral Oncol 34:304–308
11. Perera M, Al-Hebshi NN, Speicher DJ, Perera I, Johnson NW (2016) Emerging role of bacteria in oral carcinogenesis: a review with special reference to perio-pathogenic bacteria. J Oral Microbiol 8:32762. https://doi.org/10.3402/jom.v8.32762
12. Su Mun L, Wye Lum S, Kong Yuiin Sze G, Hock Yoong C, Ching Yung K, Kah Lok L, Gopinath D (2021) Association of microbiome with oral squamous cell carcinoma: a systematic review of the metagenomic studies. Int J Environ Res Public Health 18. https://doi.org/10.3390/ijerph18147224
13. Mima K, Nishihara R, Qian ZR, Cao Y, Sukawa Y, Nowak JA, Yang J, Dou R, Masugi Y, Song M, Kostic AD, Giannakis M, Bullman S, Milner DA, Baba H, Giovannucci EL, Garraway LA, Freeman GJ, Dranoff G, Garrett WS, Huttenhower C, Meyerson M, Meyerhardt JA, Chan AT, Fuchs CS, Ogino S (2016) *Fusobacterium nucleatum* in colorectal carcinoma tissue and patient prognosis. Gut 65:1973–1980. https://doi.org/10.1136/gutjnl-2015-310101
14. Nosho K, Sukawa Y, Adachi Y, Ito M, Mitsuhashi K, Kurihara H, Kanno S, Yamamoto I, Ishigami K, Igarashi H, Maruyama R, Imai K, Yamamoto H, Shinomura Y (2016) Association of *Fusobacterium nucleatum* with immunity and molecular alterations in colorectal cancer. World J Gastroenterol 22:557–566. https://doi.org/10.3748/wjg.v22.i2.557
15. Mangani S, Kremmydas S, Karamanos NK (2025) Mimicking the complexity of solid tumors: how spheroids could advance cancer preclinical transformative approaches. Cancers (Basel) 17. https://doi.org/10.3390/cancers17071161
16. Galeano Nino JL, Wu H, LaCourse KD, Kempchinsky AG, Baryiames A, Barber B, Futran N, Houlton J, Sather C, Sicinska E, Taylor A, Minot SS, Johnston CD, Bullman S (2022) Effect of the intratumoral microbiota on spatial and cellular heterogeneity in cancer. Nature 611:810–817. https://doi.org/10.1038/s41586-022-05435-0
17. Britton TA, Lee JH, Chang C, Bhat AH, Chen YW, Mohammed Ali R, Wu C, Das A, Ton-That H (2025) Inactivation of the *Fusobacterium nucleatum* Rnf complex reduces FadA-mediated amyloid formation and tumor development. MBio 16:e0103225. https://doi.org/10.1128/mbio.01032-25

Chapter 15

Fluorescence and Electron Microscopy of *Fusobacterium nucleatum*-Induced Autophagy

Yimin Chen and Hung Ton-That

Abstract

Fusobacterium nucleatum, an opportunistic oral anaerobe, has emerged as a key microbial contributor of colorectal cancer (CRC) progression and chemoresistance through its modulation of host cellular pathways. Autophagy—a conserved degradation process—has been identified as a central mechanism by which *F. nucleatum* enhances tumor cell survival. Here, we describe a standardized platform for detecting and quantifying *F. nucleatum*-induced autophagy in CRC cells using immunofluorescence microscopy with an mRFP-GFP-LC3 reporter to measure autophagic flux, complemented by ultrastructural analysis of autophagosomes via thin-section transmission electron microscopy. This protocol provides a robust and reproducible framework for investigating host-microbe interactions in CRC and can be readily adapted to evaluate autophagy induction by other bacterial species.

Key words *Fusobacterium nucleatum*, Colorectal cancer, Autophagy, Electron microscopy, Immunofluorescence microscopy, mRFP-GFP-LC3, Autophagosome

1 Introduction

Autophagy is an evolutionarily conserved lysosomal degradation pathway that maintains cellular homeostasis through the turnover of cytoplasmic components, damaged organelles, and intracellular pathogens [1, 2]. This dynamic process begins with the formation of double-membrane autophagosomes that encapsulate cytoplasmic cargo and proceeds through their maturation and fusion with lysosomes to generate autolysosomes, where contents are degraded. Precise monitoring of autophagic flux—the entire continuum from autophagosome biogenesis to cargo degradation—is essential to distinguish true autophagy activation from incomplete flux or vesicle accumulation due to downstream blockage [2]. Understanding these distinctions is particularly important when examining how external biological stimuli, such as microbial pathogens, modulate autophagy within host cells.

Hung Ton-That (ed.), *Fusobacterium nucleatum: Methods and Protocols*, Methods in Molecular Biology, vol. 3055, https://doi.org/10.1007/978-1-0716-5388-3_15,

Fusobacterium nucleatum, a Gram-negative obligate oral anaerobe frequently enriched in the colorectal cancer (CRC) microenvironment, has emerged as a microbiome-derived modulator of autophagy [3, 4]. *F. nucleatum* not only adheres to and invades epithelial and cancer cells but also reprograms host signaling pathways to subvert immune surveillance, promote tumor progression, and induce chemoresistance [5]. Notably, this organism has been shown to stimulate autophagosome formation while potentially impairing terminal autophagic degradation, suggesting that it may exploit dysregulated autophagy to facilitate intracellular persistence and enhance survival [3, 6, 7].

To characterize *F. nucleatum*-induced autophagy with both functional and structural resolution, we present detailed protocols for two complementary microscopic approaches. First, autophagic flux is quantified using the tandem mRFP-GFP-LC3 reporter, which distinguishes GFP^+/RFP^+ autophagosomes from acidified RFP-only autolysosomes, enabling compartment-specific assessment of autophagy dynamics [8]. Second, thin-section transmission electron microscopy (TEM) offers ultrastructural validation of autophagosomes and their spatial relationship to intracellular bacteria [9, 10]. Together, these methods provide a robust and integrated framework for defining microbial modulation of autophagy in CRC cells and can be readily adapted for other host-microbe systems.

2 Materials

All solutions should be prepared with sterile molecular grade water.

2.1 Fluorescence Imaging Using Tandem mRFP-GFP-LC3 Reporter

2.1.1 Coverslip Preparation

1. Coverslip (12 mm diameter).
2. 0.01% poly-L-lysine solution prepared from lyophilized poly-L-lysine hydrobromide in water (*see* **Note 1**).
3. Sterile water.
4. 6-well or 12-well tissue culture plates.
5. Sterile forceps.

2.1.2 Cell Preparation for Transfection

1. Mammalian CRC cell line of interest (e.g., HT29, HCT116, or DLD1).
2. Complete growth medium: McCoy's 5A medium supplemented with 10% FBS, L-glutamine, and 1% penicillin-streptomycin.
3. RPMI-1640-modified serum-free medium containing 4.5 g/L of D-glucose, 2.383 g/L of HEPES buffer, L-glutamine, 1.5 g/L of sodium bicarbonate, and 110 mg/L of sodium pyruvate.

4. Trypsin-EDTA solution for cell passaging.
5. Hemocytometer.
6. Cell culture incubator set to 37 °C with 5% CO_2.

2.1.3 Transfection of mRFP-GFP-LC3 Plasmid

1. Lipofectamine 3000 transfection reagent.
2. RPMI-1640-modified serum-free medium containing 4.5 g/L of D-glucose, 2.383 g/L of HEPES buffer, L-glutamine, 1.5 g/L of sodium bicarbonate, and 110 mg/L of sodium pyruvate.
3. mRFP-GFP-LC3 plasmid (e.g., ptfLC3 and Addgene).
4. 50 mg/mL of G418 in HEPES buffer.
5. Cell culture incubator set to 37 °C with 5% CO_2.

2.1.4 F. nucleatum Infection of CRC Cell

1. *F. nucleatum* CW1—a derivative of *F. nucleatum* ATCC 23726 [11].
2. TSP: Mix 3 g of BD Bacto Tryptic Soy Broth and 1 g of BD Gibco peptone in 100 mL of water.
3. TSPC medium: Mix 20 mL of TSP medium with 0.1 mL of 10% cysteine.
4. 1.5-mL microcentrifuge tubes.
5. Phosphate buffered saline (PBS), pH 7.4.
6. RPMI-1640-modified serum-free medium.
7. Trypsin-EDTA solution for cell passaging.
8. 6-well or 12-well tissue culture plates.
9. Cell culture incubator set to 37 °C with 5% CO_2.
10. Anaerobic chamber with appropriate gas mixture (*see* **Note 2**).

2.1.5 Fixation and Mounting

1. Ice-cold PBS, pH 7.4.
2. Dumont angled tweezers.
3. Fluorescence microscope equipped with appropriate GFP and RFP filters.
4. Ice-cold acetone (*see* **Note 4**).
5. 1 μg/mL of DAPI solution (optional for nuclear staining).
6. Microscope slides (25 mm × 75 mm × 1 mm).
7. Nail polish for sealing coverslips.

2.1.6 Fluorescence Imaging

1. Fluorescence microscope equipped with appropriate GFP and RFP filters.
2. Confocal fluorescence microscope equipped with 488-nm (GFP) and 561-nm (RFP) lasers.

3. Appropriate filters for GFP and RFP fluorescence detection (*see* **Note 5**).
4. Image acquisition software (e.g., Keyence BZ-X) (*see* **Note 6**).

2.2 Thin-Section Transmission Electron Microscopy (TEM)

2.2.1 Cell Culture and Infection

1. Mammalian cell lines (e.g., HT29, HCT116, or DLD1).
2. Complete McCoy's 5A medium with 10% FBS, 1% penicillin-streptomycin, and (+) L-glutamine.
3. RPMI-1640-modified serum-free medium containing 4.5 g/L of D-glucose, 2.383 g/L of HEPES buffer, L-glutamine, 1.5 g/L of sodium bicarbonate, and 110 mg/L of sodium pyruvate.
4. 0.05% Trypsin-EDTA.
5. PBS, pH 7.4.
6. *F. nucleatum* CW1.
7. TSPC.
8. Hemocytometer.
9. Anaerobic chamber with appropriate gas mixture.
10. 10-cm tissue culture dishes.
11. Sterile cell scrapers or razor blades.
12. 15-mL conical centrifuge tubes.

2.2.2 Fixation and Staining

1. 4.5% paraformaldehyde (electron microscopy grade).
2. 0.2 M HEPES, pH 7.4.
3. 0.1 M phosphate buffer, pH 7.4.
4. Cell scraper.
5. Ice bucket or cold room access for chilled fixation steps.
6. Rocking platform for gentle agitation.
7. 2% osmium tetroxide (aqueous) (*see* **Note 7**).
8. 1.5% potassium ferricyanide (freshly prepared).
9. 2% aqueous uranyl acetate.
10. Milli-Q or double-distilled ice-cold water.
11. Aluminum foil (for light protection of uranyl acetate).

2.2.3 Dehydration and Resin Infiltration

1. Ice-cold ethanol series: 50%, 70%, 85%, 95%, and 100%.
2. 100% anhydrous ethanol (fresh or stored over 3 Å molecular sieves).
3. Ice-cold acetone, anhydrous (optional substitute for propylene oxide) (*see* **Note 8**).
4. 1.5-mL Eppendorf or glass vials for solvent exchange.

2.2.4 Embedding and Polymerization

1. Spurr's resin kit (e.g., Electron Microscopy Sciences).
2. Beem embedding capsules.
3. Embedding molds (flat or pyramid shape).
4. Oven or vacuum oven set to 65 °C.
5. Vacuum desiccator (optional but recommended).

2.2.5 Ultrathin Sectioning

1. Ultramicrotome with diamond knife.
2. Fresh-shaped glass knife.
3. 200-mesh copper grids (square or hexagonal mesh).
4. Grid storage box.
5. Fine-tipped forceps.

2.2.6 Post-Staining

1. U2% aqueous uranyl acetate (for post-section staining).
2. 0.3% lead citrate (*see* **Note 9**).
3. TEM grid staining setup (watch glasses, filter paper, CO_2 absorber for lead citrate).

2.2.7 TEM Imaging

1. Transmission electron microscope (e.g., TF20 TEM).
2. TEM imaging software (e.g., SerialEM automatic data acquisition Version 4.0.0).

3 Methods

3.1 Fluorescence Imaging Using Tandem mRFP-GFP-LC3 Reporter

3.1.1 Coverslip Preparation

1. Place sterile coverslips into the wells of a six-well plate using sterile forceps.
2. Add enough 0.01% poly-L-lysine solution to fully cover each coverslip (*see* **Note 10**).
3. Incubate at room temperature for 1 h.
4. Aspirate the poly-L-lysine and wash once with sterile water.
5. Allow coverslips to dry completely inside a biosafety cabinet.

3.1.2 Cell Preparation for Transfection

1. Culture CRC cells in complete McCoy's 5A medium supplemented with 10% FBS, L-glutamine and 1% penicillin-streptomycin at 37 °C in a 5% CO_2 incubator with until they reach 70–80% confluency.
2. Aspirate the culture medium and gently wash cells once with pre-warmed PBS.
3. Trypsinize the cells, and then resuspend them in complete RPMI-1640-modified serum-free medium containing 4.5 g/L of D-glucose, 2.383 g/L of HEPES buffer, L-glutamine, 1.5 g/L of sodium bicarbonate, and 110 mg/L of sodium pyruvate.

4. Determine cell density using a hemocytometer.
5. Seed 5 × 10^5 CRC cells onto poly-L-lysine-coated coverslips prepared in Subheading 3.1.1, placing them in a six-well plate at approximately 50% confluency (*see* **Note 11**).
6. Incubate the cells overnight at 37 °C in a 5% CO_2 incubator.

3.1.3 Transfection of mRFP-GFP-LC3 Plasmid

1. Dilute 5 μL of Lipofectamine 3000 in 250 μL of RPMI-1640-modified serum-free medium.
2. Dilute 1–2 μg of mRFP-GFP-LC3 plasmid DNA in 100 μL of the same serum-free RPMI-1640 medium.
3. Combine the diluted Lipofectamine and plasmid solutions, and then adjust the total volume to 1 mL with the same serum-free medium (*see* **Note 12**).
4. Incubate the DNA-lipid complexes for 30 min at room temperature (*see* **Note 13**).
5. Add the complexes dropwise onto CRC cells prepared in Subheading 3.1.2 in complete medium.
6. Incubate the cells for 48 h at 37 °C.
7. Selected transfected CRC cells with G418 (1000 μg/mL) for 1 week to obtain a fully fluorescent, stably expressing population (*see* **Note 14**). CRC cells expressing mRFP-GFP-LC3 may be cryopreserved for future experiments.

3.1.4 *F. nucleatum* Infection of CRC Cells

1. Grow the selected CRC cells stably expressing mRFP-GFP-LC3 (prepared in Subheading 3.1.3) on coverslips 1 day prior to infection.
2. Culture *F. nucleatum* anaerobically in TSPC medium at 37 °C overnight.
3. On the day of infection, dilute the overnight culture to an OD_{600} of 0.1, and incubate for roughly 4 h or until it reaches an OD_{600} of 0.5.
4. Harvest bacteria by centrifugation at 6500× *g* for 5 min, and wash once with sterile PBS.
5. Resuspend the bacterial pellet in antibiotic-free RPMI-1640 serum-free medium to achieve the desired multiplicity of infection (MOI), typically 500:1 for autophagy assays (*see* **Note 15**).
6. Aspirate the culture medium from CRC cells on coverslips in a six-well plate.
7. Wash CRC cells twice with PBS, and replace the culture medium with the prepared bacterial suspension.
8. Incubate CRC cells with fusobacteria for 4.5 h at 37 °C in a 5% CO_2 incubator, including uninfected control wells.

9. Following infection, aspirate the bacterial suspension, and gently wash the cells three times with warm PBS.
10. Replace PBS with fresh complete medium and proceed immediately to fixation.

3.1.5 Fixation and Mounting

1. Gently wash CRC cells twice with ice-cold PBS.
2. Fix cells with ice-cold acetone for 1 min at room temperature (*see* **Note 16**).
3. Wash three times with ice-cold PBS.
4. Using Dumont angled tweezers, carefully retrieve the coverslip (with adherent CRC cells) from the six-well culture plate.
5. Apply DAPI to a microscope slide and mount the coverslip onto the slide.
6. Scal thc coverslip edges with clear nail polish and allow to dry for 5 min. Store slides at 4 °C, protected from light. Slides may be stored at −20 °C. Before reimaging, inspect the seal and reapply nail polish as needed.

3.1.6 Fluorescence Imaging

1. Acquire images on a fluorescence microscope using the following channels:
   (a) GFP/AlexaFluor 488: Ex/Em: 490/525 nm (green).
   (b) DAPI: Ex/Em 350/470 nm (blue).
   (c) FM 4-64: Ex/Em 515/640 nm (red).
2. Maintain identical exposure times, gain, and illumination settings for all samples within the same experiment to ensure comparability.
3. Focus using the DAPI channel to identify nuclei, and then sequentially capture images in the GFP/Alexa Fluor 488 and FM 4-64 channels to minimize spectral overlap.
4. Acquire images at 20× or 40× magnification, collecting at least five randomly selected fields of view per sample.
5. Save all images in a lossless format (e.g., TIFF) for downstream quantitative analysis.
6. Perform basic image processing—such as background subtraction and channel merging—using ImageJ or equivalent image-analysis software (*see* Fig. 1a for an example).

3.2 Thin-Section Transmission Electron Microscopy (TEM)

3.2.1 Cell Culture and Infection

1. Prepare *F. nucleatum* cultures as described in **steps 1–3** of Subheading 3.1.4. Harvest bacteria, and resuspend the pellet in in antibiotic-free RPMI mod medium to achieve the desired multiplicity of infection (typically MOI 500:1 for autophagy assay).

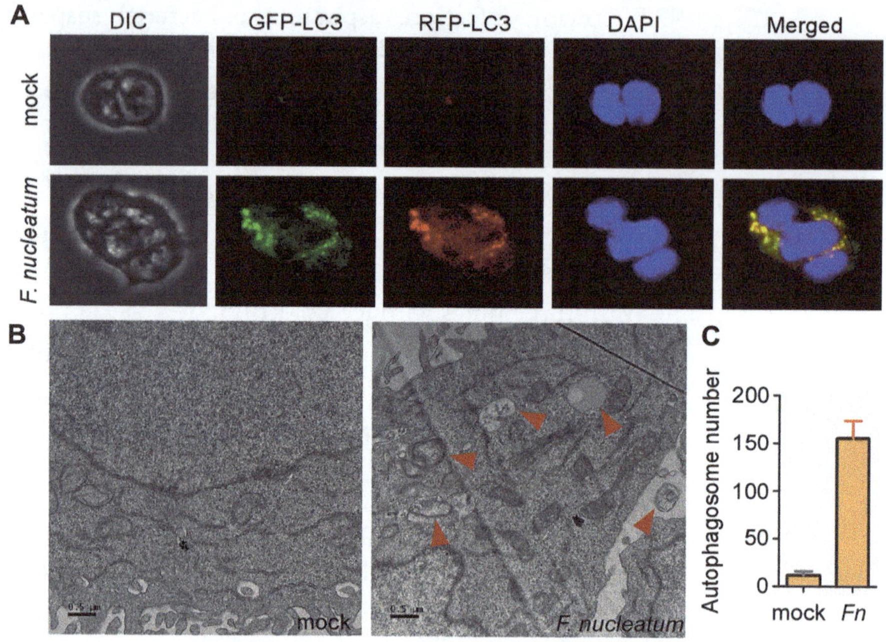

Fig. 1 Fluorescence and electron microscopy of autophagosomes. (**a**) *F. nucleatum* infection induces autophagosome formation in HCT116 CRC cells. Representative confocal images of mock- or *F. nucleatum*-treated cells expressing the mRFP-GFP-LC3 reporter are shown. Infected cells exhibit a pronounced increase in GFP-LC3 (green) and RFP-LC3 (red) puncta, with abundant yellow puncta in merged images, indicating accumulation of autophagosomes. Nuclei are counterstained with DAPI (blue). (**b**, **c**) Shown are representative thin-section TEM images of mock- or *F. nucleatum*-treated HCT116 cells. Red arrowheads denote autophagosomes. Quantification of autophagosome numbers in b is presented in c. Scale bars, 0.5 μm

2. Prepare 1.5×10^6 CRC cells (HCT116 or HT29), seed them into a 6-cm culture plate containing complete McCoy's 5A medium, and grow overnight at 37 °C in a 5% CO_2 incubator.
3. Aspirate the culture medium and gently wash CRC cells twice with PBS.
4. Replace the culture medium with antibiotic-free, serum-free RPMI-1640-modified medium.
5. Infect CRC cells by adding the fusobacterial suspension (*see* **step 1**) and incubating for 4.5 h at 37 °C in a 5% CO_2 incubator. Include uninfected plates as negative controls.

3.2.2 Fixation and Staining

1. At 4.5-h postinfection, add warm (37 °C) 4.5% paraformaldehyde in 0.2 M HEPES (pH 7.4) directly to the culture plate (*see* Subheading 3.2.1, **step 5**) and incubate for 15–30 min at room temperature (*see* **Note 17**).
2. Gently scrape the cells, and collect them in Eppendorf tubes by centrifugation at 500× *g* for 5 min, followed by a brief spin at 12,000× *g* for 2 min.

3. Place the tubes on ice, and continue fixation by incubating the pellets overnight in 4.5% paraformaldehyde in 0.2 M HEPES (pH 7.4) on ice (*see* **Note 18**).
4. Wash the pellets five times with 1 mL of ice-cold 0.1 M HEPES (pH 7.4), 5 min each wash, keeping samples on ice throughout.
5. Post-fix the pellets in a solution of 2% osmium tetroxide and 1.5% potassium ferricyanide prepared in 0.1 M phosphate buffer. Incubate on ice for 1 h in the dark (*see* **Note 19**).
6. Wash the samples five times with 1 mL of ice-cold Milli-Q water, 5 min per wash.
7. Stain the pellets with 2% aqueous uranyl acetate on ice for at least 2 h or overnight, keeping the samples wrapped in aluminum foil to protect them from light.
8. Wash the samples again five times with ice-cold water, 5 min per wash, before proceeding to dehydration.

3.2.3 Dehydration and Resin Infiltration

1. Dehydrate samples through a graded, ice-cold ethanol series (50%, 70%, 85%, 95%, and 100%), 7 min at each step.
2. Incubate pellets in 100% ethanol (freshly opened or stored over 3 Å molecular sieves) twice for 15 min each, followed by a 20-min incubation in acetone (*see* **Note 20**).
3. Transfer samples to 33% resin in anhydrous acetone and incubate for 30–60 min.
4. Infiltrate in 66% resin in acetone overnight (not exceeding 16 h).
5. Transfer samples to 100% resin and incubate for 2–3 h at room temperature.

3.2.4 Embedding and Polymerization

1. Transfer samples to Beem capsules filled with fresh 100% Spurr's resin (*see* **Note 21**).
2. Embed the pellets in fresh 100% resin in Beem capsules and polymerize at 65 °C for 48 h, preferably under vacuum to minimize bubble formation.

3.2.5 Ultrathin Sectioning

1. Trim resin blocks into a rectangular shape using a freshly shaped glass knife.
2. Cut ~50-nm ultrathin sections using a diamond knife on an ultramicrotome (*see* **Note 22**).
3. Collect sections by using fine-tipped forceps onto 200-mesh copper grids.

3.2.6 Post-Staining

1. Stain grids with 2% uranyl acetate in water for 15 min.
2. Rinse thoroughly with water and blot dry.
3. Stain with 0.3% lead citrate for 3 min in a CO_2-free environment (*see* **Note 23**).
4. Rinse grids thoroughly again and air dry.

3.2.7 TEM Imaging

1. Load the dried grids into a transmission electron microscope (e.g., TF20).
2. Image samples at accelerating voltage of 80–120 kV (*see* **Note 24**).
3. Acquire digital images of infected and uninfected cells using appropriate TEM software (*see* **Note 25**).
4. Analyze autophagosomes, phagosomes, and cellular ultrastructure to assess response to *F. nucleatum* infection (*see* Fig. 1b, c for an example).

4 Notes

1. Poly-L-lysine coating is critical for cell adhesion. Ensure that the 12-mm coverslips are properly coated with 0.01% poly-L-lysine. Uneven coating or inadequate incubation can lead to poor attachment and substantial cell loss during fixation or washing. Coverslips should be completely air-dried in a sterile environment prior to use.
2. *F. nucleatum* exhibits pronounced sensitivity to oxygen, and transient exposure to ambient conditions can substantially diminish viability and reduce infection efficiency. Throughout experimental workflows, materials such as tubes, plates, and culture media are maintained under anaerobic conditions and secured to preserve the low-oxygen environment upon removal. Media are typically equilibrated anaerobically in advance to ensure consistent bacterial performance in downstream assays.
3. Transfection efficiency varies by CRC cell lines. HCT116 and HT29 generally show higher Lipofectamine 3000 transfection rates, whereas DLD1 often requires optimization (e.g., adjusting the DNA:reagent ratio). Minimize transfection-induced autophagy by reducing exposure time to the reagent and limiting serum-free RPMI incubation to $\leq$48 h.
4. Acetone fixation is rapid but harsh. Fixation longer than 10 min can distort subcellular structures and diminish fluorescence intensity.
5. The mRFP-GFP-LC3 reporter is pH-sensitive. GFP signal is quenched in acidic lysosomes, while RFP remains stable.

However, prolonged illumination can photobleach RFP and lead to misinterpretation. Acquire images promptly and avoid overexposure.

6. Verify imaging system calibration before acquisition. Confirm correct laser alignment, camera focus, and software settings; even minor adjustments in gain or exposure can influence quantitative analysis.
7. Strict temperature and timing control are essential for EM fixation. Paraformaldehyde should be freshly diluted and chilled. Osmium tetroxide (2%) must be prepared and handled in a fume hood. Incubation longer than 1 h can damage membranes. Keep samples on ice unless otherwise noted.
8. Use molecular-grade, pre-chilled ethanol and acetone. Residual water during dehydration interferes with resin infiltration and causes sectioning artifacts. Fresh desiccant improves dehydration efficiency. Each dehydration step should last at least 10 min for optimal resin penetration.
9. Ultramicrotomy requires consistent section thickness. Irregular thickness leads to blurred TEM images and uneven contrast. Lead citrate is highly CO_2-sensitive and easily precipitates; stain in a CO_2-free chamber to avoid black artifacts.
10. Ensure coverslips are fully dry after poly-L-lysine coating. Residual solution results in poor cell adhesion and uneven growth.
11. Coverslips can be sterilized with ethanol and flame-dried. Avoid oversaturating with cells, as high confluency reduces transfection efficiency and can alter autophagic responses.
12. Optimize the Lipofectamine-to-DNA ratio for each cell line. Excess reagent causes cytotoxicity, while insufficient reagent reduces transfection efficiency. Perform a small gradient test for each new cell line.
13. Use pre-warmed, supplemented RPMI. Cold or improperly buffered medium compromises complex formation and may reduce infection efficiency.
14. G418 selection should be introduced gradually. Although 1000 μg/mL is effective for most CRC cell lines (HCT116, HT29, DLD1), titration is recommended. Monitor fluorescence daily to assess selection progress.
15. An MOI of 500:1 is suitable for autophagy assays but may cause cytotoxicity in sensitive lines. Consider testing MOIs such as 50:1 or performing time-course experiments to optimize viability and signal.
16. Acetone fixation is ideal for rapid immunofluorescence of cytosolic or nuclear proteins such as LC3, as it simultaneously fixes

and permeabilizes within 1–2 min. However, its harshness makes it unsuitable for membrane-associated targets or experiments requiring fine ultrastructural preservation.

17. Fixing exactly 15 min postinfection captures early autophagic events. Delays can diminish signal or shift detection toward late-stage structures.
18. Keep samples on ice immediately after centrifugation. Aldehyde fixation should proceed on ice for a minimum of 2 h and may be extended up to overnight without issue.
19. Osmium–ferricyanide fixation increases contrast but is highly toxic. Always prepare and handle this mixture in a fume hood. Maintain cold, dark incubation to maximize staining quality.
20. Gradual dehydration prevents cell shrinkage. Ensure ethanol is either freshly opened or stored over sieves. Avoid introducing water into the resin, as even trace amounts impede polymerization.
21. Vacuum polymerization reduces bubble formation in resin blocks. Incubation at 65 °C for at least 48 h is optimal. Verify block hardness before trimming.
22. Ultrathin sectioning (~50 nm) requires a clean, sharp diamond knife. A dull blade causes compression, chatter marks, and uneven sections.
23. Lead citrate must be used in a CO_2-free environment. Perform staining in a sealed chamber containing NaOH pellets to prevent carbonate precipitation that obscures TEM images.
24. Maintain consistent imaging voltage (80–120 kV) for reproducibility. Keep the electron beam centered and limit exposure to reduce drift and charging artifacts.
25. Accurate TF20 alignment is essential. Inadequate wobble correction compromises focus on high magnification. Carefully perform Z-height and eucentric height calibration prior to imaging.

Conflict of Interest

The authors have no conflicts of interest to declare that are relevant to the content of this chapter.

References

1. Mizushima N, Komatsu M (2011) Autophagy: renovation of cells and tissues. Cell 147:728–741. https://doi.org/10.1016/j.cell.2011.10.026
2. Mizushima N, Yoshimori T, Levine B (2010) Methods in mammalian autophagy research. Cell 140:313–326. https://doi.org/10.1016/j.cell.2010.01.028

3. Yu T, Guo F, Yu Y, Sun T, Ma D, Han J, Qian Y, Kryczek I, Sun D, Nagarsheth N, Chen Y, Chen H, Hong J, Zou W, Fang JY (2017) Fusobacterium nucleatum promotes chemoresistance to colorectal cancer by modulating autophagy. Cell 170:548–563 e16. https://doi.org/10.1016/j.cell.2017.07.008
4. Kostic AD, Chun E, Robertson L, Glickman JN, Gallini CA, Michaud M, Clancy TE, Chung DC, Lochhead P, Hold GL, El-Omar EM, Brenner D, Fuchs CS, Meyerson M, Garrett WS (2013) Fusobacterium nucleatum potentiates intestinal tumorigenesis and modulates the tumor-immune microenvironment. Cell Host Microbe 14:207–215. https://doi.org/10.1016/j.chom.2013.07.007
5. Brennan CA, Garrett WS (2019) Fusobacterium nucleatum - symbiont, opportunist and oncobacterium. Nat Rev Microbiol 17:156–166. https://doi.org/10.1038/s41579-018-0129-6
6. Bullman S, Pedamallu CS, Sicinska E, Clancy TE, Zhang X, Cai D, Neuberg D, Huang K, Guevara F, Nelson T, Chipashvili O, Hagan T, Walker M, Ramachandran A, Diosdado B, Serna G, Mulet N, Landolfi S, Ramon YCS, Fasani R, Aguirre AJ, Ng K, Elez E, Ogino S, Tabernero J, Fuchs CS, Hahn WC, Nuciforo P, Meyerson M (2017) Analysis of Fusobacterium persistence and antibiotic response in colorectal cancer. Science 358:1443–1448. https://doi.org/10.1126/science.aal5240
7. Duan C, Tang X, Wang W, Qian W, Fu X, Deng X, Zhou S, Han C, Hou X (2021) Lactobacillus rhamnosus attenuates intestinal inflammation induced by Fusobacterium nucleatum infection by restoring the autophagic flux. Int J Mol Med 47:125–136. https://doi.org/10.3892/ijmm.2020.4780
8. Kimura S, Noda T, Yoshimori T (2007) Dissection of the autophagosome maturation process by a novel reporter protein, tandem fluorescent-tagged LC3. Autophagy 3:452–460. https://doi.org/10.4161/auto.4451
9. Wyffels JT (2001) Principles and techniques of electron microscopy: biological applications, fourth edition, by M. A Hayat. Microsc Microanal 7:66
10. Massover WH (2011) New and unconventional approaches for advancing resolution in biological transmission electron microscopy by improving macromolecular specimen preparation and preservation. Micron 42:141–151. https://doi.org/10.1016/j.micron.2010.05.006
11. Wu C, Al Mamun AAM, Luong TT, Hu B, Gu J, Lee JH, D'Amore M, Das A, Ton-That H (2018) Forward genetic dissection of biofilm development by Fusobacterium nucleatum: novel functions of cell division proteins FtsX and EnvC. mBio 9. https://doi.org/10.1128/mBio.00360-18

Chapter 16

Mass Spectrometry-Based Untargeted Metabolite Profiling to Discover Molecular Consequences of Mammalian Cell Infection with *Fusobacterium nucleatum*

Qiuying Chen, Huiyong Cheng, and Steven S. Gross

Abstract

Untargeted metabolite profiling, aka *metabolomics*, offers a powerful technology to potentially reveal fundamental molecular mechanisms whereby a perturbation in cell metabolism leads to an alteration in cell growth or physiological maintenance. Notably, this discovery approach does not require any preconceived notion of underlying molecular mechanisms, affording an opportunity to effectively shine a molecular mechanistic spotlight in a direction that an informed investigator may not have thought to consider. *Fusobacterium nucleatum* colonization has been linked to the initiation and promotion of the colorectal cancer and other pathologies. Although *F. nucleatum* infection can modulate mammalian cell signaling pathways involved in inflammation and host defense, the broad impact of *F. nucleatum* on host cell metabolism is poorly defined. Here, we describe the application of untargeted LC/MS-based profiling of cell metabolites as a tool to comprehensively assess the impact of *F. nucleatum* infection on mammalian cell metabolism. Untargeted metabolite profiling offers an unprecedented opportunity for the generation of novel molecular mechanistic hypotheses for subsequent confirmation by targeted orthologous investigations.

Key words Metabolomics, Untargeted metabolite profiling, Liquid chromatography-mass spectrometry, LC/MS, *Fusobacterium nucleatum*, Gram-negative bacteria, Spheroids

1 Introduction

Gene expression technologies (e.g., transcriptomics and proteomics) have been widely employed for unbiased discovery of molecular consequences that arise from gene mutations, drug treatments, and disease states in mammalian cells. Despite the utility of these omics discovery tools, small molecule metabolites comprise the most proximal functional effector of perturbed mammalian gene, transcript, and protein expression. Consequently, knowledge of the ever-changing small molecule composition of mammalian cells, i.e., the *metabolome*, offers the greatest potential insight into the

Hung Ton-That (ed.), *Fusobacterium nucleatum: Methods and Protocols*, Methods in Molecular Biology, vol. 3055,
https://doi.org/10.1007/978-1-0716-5388-3_16,

moment-to-moment status of mammalian cell function and/or dysfunction. A major challenge to obtaining a comprehensive inventory of small molecule metabolites in cells is that, unlike the case for transcripts and proteins, the mammalian cell metabolome is largely undefined and cannot be predicted from knowledge of the mammalian genome. In addition to not knowing the identity of all endogenously-derived metabolites in mammalian cells, a further complication is that cellular metabolites also derive from exogenous environmental sources—including the air we breathe, foods we eat, and our unique microbiomes. Given that *F. nucleatum* is well-documented to perturb mammalian cell function and has been implicated in conditions that include worsened colorectal cancer outcomes [1, 2], we describe a state-of the-art strategy for untargeted metabolite profiling that can be applied to uncover fundamental metabolic consequences of *F. nucleatum* infection that can contribute to mammalian cell pathobiology.

2 Materials

2.1 Cancer Colonocyte Cell Culture and F. nucleatum Infection

1. Spheroids of colorectal cancer (CRC) cells (e.g., HCT116 and HT29), CRC organoids, or CRC tumors.
2. *F. nucleatum* strains (e.g., ATCC 23726 and ATCC 25586).
3. *E. coli* (control).
4. Dulbecco's Modified Eagle's Medium (DMEM).
5. Fetal bovine serum (FBS).

2.2 Preparation of Cell Extracts for Metabolite Analysis by LC/MS

1. Phosphate-buffered saline (PBS).
2. Vacuum aspirator.
3. MS-grade methanol (MeOH).
4. Double-distilled H_2O (ddH_2O) or Millipore 18 meg Ω resistance H_2O.
5. 2.0-mL Eppendorf tubes.
6. −20 °C freezer.
7. 2-mL Eppendorf safe-lock round bottom tubes for cell disruption by bead-beating.
8. Qiagen TissueLyser III (bead-beater).
9. Stainless-steel beads.
10. Refrigerated microfuge.
11. A SpeedVac system.

2.3 Protein Pellet Solubilization and Quantification for Normalization of LC/MS Sample Input

1. 0.2 M NaOH.
2. Qiagen TissueLyser III (bead-beater).
3. Stainless-steel beads.
4. Heat-blocks.
5. Vortexers.
6. Eppendorf tubes.
7. Protein Assay Kit (e.g., BioRad DC).

2.4 Reconstitution of Dried Metabolite Extracts for LC/MS-Based Metabolite Profiling

1. ddH_2O.
2. LC-MS grade acetonitrile.
3. LC/MS grade isopropanol.
4. Microfuge.
5. Spectrophotometric microplate reader (wavelength 650–750 nm range).
6. OmniTrace Ultra™ acetic acid.
7. Ammonium acetate.
8. Ammonium formate.
9. Ethylenediaminetetraacetic acid.
10. Agilent deactivator additive.
11. LC autosampler.
12. Agilent ES-TOF Biopolymer Analysis Reference Mass Standards Kit.
13. Agilent Model 1290 Infinity II liquid chromatography system.
14. Agilent 6550 iFunnel time-of-flight MS analyzer.
15. Cogent Diamond Hydride Dursan column (4 μm, 100A, 150 mm x 21 mm ID).
16. Agilent ZORBAX Eclipse Plus C18 column (1.8 μm, 100 × 2.1 mm ID).
17. Vortex, ice, vacuum suction pipettes, and transfer pipettes.

3 Methods

3.1 Cancer Colonocyte Cell Culture and Fuso Infection

1. Prepare human colorectal cell (CRC) spheroids infected by *F. nucleatum* (*see* Chapter 14 for details). Alternatively, CRC organoids in vitro and isolated CRC tumors in vivo can serve as cell sources to assess fundamental metabolic perturbations that arise from *F. nucleatum* infection.
2. On the day prior to metabolite extraction, cell culture medium is replaced with a modified medium that is devoid of antibiotics and contains extensively dialyzed FBS along with either

F. nucleatum at the desired MOI, a control bacterial strain (e.g., *E. coli*), or an equivalent volume of aseptic culture medium.

3.2 Preparation of Cell Extracts for Metabolite Analysis by LC/MS

1. Remove media from spheroid cultures and wash twice with 2 mL of PBS, using vacuum aspiration to take off as much PBS as possible after each cell wash. Isolated tumors may similarly be PBS-washed in preparation for metabolite extraction.
2. Perform a rapid final rinse of cells or tumors with 2 mL of cold ddH_2O to remove as much residual salt as possible (*see* **Note 1**).
3. Rapidly quench colonocyte metabolism by addition of 400 μL of ultra-cold (−70 °C to −80 °C) 80% MeOH/ddH_2O to cells in culture wells, or tumors in 2.0-mL Eppendorf bead-beater tubes. For cells in culture, scrape-harvest into 2.0-mL Eppendorf bead-beater tubes and quantitatively recover extracts from culture wells by washing wells two additional times with 150 μL of MeOH/ddH_2O—pool all transfers.
4. Incubate Eppendorf tubes containing scrape-harvested cells or tumors in a −20 °C freezer for 10 min. Eppendorf tubes with methanolic cell extracts can either be processed as described below, or preferably, be stored at −80 °C with further processing to completion on the day of LC/MS analysis.
5. For cell disruption and metabolite extraction, add a single stainless-steel bead to Eppendorf tubes and bead-beat in a Qiagen TissueLyser at high-speed for 45 s.
6. Incubate the resulting methanolic cell extracts at −20 °C for 10 min, followed by centrifugation at 20,800× *g* for 5 min.
7. Quantitatively transfer tissue extracts into clean Eppendorf tubes. After transfer of the initial bead-beated sample supernatant to the Eppendorf tube, add an additional 200 μL of −80 °C 80% MeOH/ddH_2O to the original cell pellets, and repeat bead-beating/centrifugation **steps 5** and **6**, twice. Pool the supernatant cell extracts and save the pellets for subsequent protein quantification (*see* **Note 2**).
8. Perform a final spin of the pooled supernatant at 20,800× *g*. Transfer the final extracted cell supernatant to a new 1.5-mL Eppendorf tube for vacuum drying in a SpeedVac (~4 h) (*see* **Note 3**).
9. Store dried cell extract samples at −80 °C until reconstitution on the day of analysis by LC/MS.

3.3 Protein Pellet Solubilization and Quantification for Normalization of LC/ MS Sample Input

1. Aliquot 100 μL of 0.2 M NaOH to cell protein pellets that remain in bead-beater tubes after metabolite extraction (*see* Subheading 3.2, **step** 7 above) (*see* **Note 4**).
2. Vigorously vortex or bead-beat the pellet to extract protein for efficient solubilization.
3. Heat tubes to 95 °C for 20 min until pellet fully dissolves and the extraction medium is clarified (*see* **Note 5**).
4. Spin down nonprotein cell debris at 20,800× *g* for 1 min.
5. Perform a BioRad DC protein assay on cell pellet extracts as described by the manufacturer's protocol.
6. Quantify the total protein content in each sample, relative to BSA standards. Use the measured protein content of samples to normalize LC/MS metabolite injection volumes to derive from an equivalent quantity of extracted cell protein.

3.4 Reconstitution of Dried Metabolite Extracts for LC/MS-Based Metabolite Profiling

1. Reconstitute dried cell extracts on the same day as LC/MS sample analyses are to be performed.
2. Add the desired volume of 70:30 acetonitrile (ACN):H_2O, vortex, with reconstitution volume normalized to original cell protein content, and incubate on ice for 10 min. Repeat with two additional cycles of vortexing and incubations on ice (*see* **Note 6**).
3. Centrifuge and transfer the solubilized cell extract supernatant to an LC autosampler vial for analysis by LC/MS (*see* **Note** 7).
4. Prior to introduction into the LC autosampler, all sample vials should be tightly capped using a crimp sealer to minimize sample losses due to evaporation. Preferentially, from a vial containing of 10 μl of sample, inject 4–6 μL for analysis by LC/MS.

3.5 LC/MS-Based Profiling of Metabolite Extracts: Data Acquisition

Analysis of cell extracts by LC/MS is performed to obtain three fundamental and distinct parameters for each recognized molecular feature: (1) an accurate mass/charge ratio (sufficient for confident molecular formula assignment), a chromatographic retention time (sufficient for matching to a retention time database of pure molecular standards), and total ion counts (sufficient for quantification of relative molecular abundance). Notably, cellular metabolites exhibit a broad range of biophysical properties, ranging from extremely hydrophilic, to extremely hydrophobic. Accordingly, maximal metabolomic coverage can be obtained by the combined use of both hydrophilic and hydrophobic chromatographic separations (i.e., using distinct column types) as a means to provide informative retention times for all observed molecular features (*see* Fig. 1 for schema). In addition to dual LC gradient pumps used for the gradient chromatography separations described below, a separate

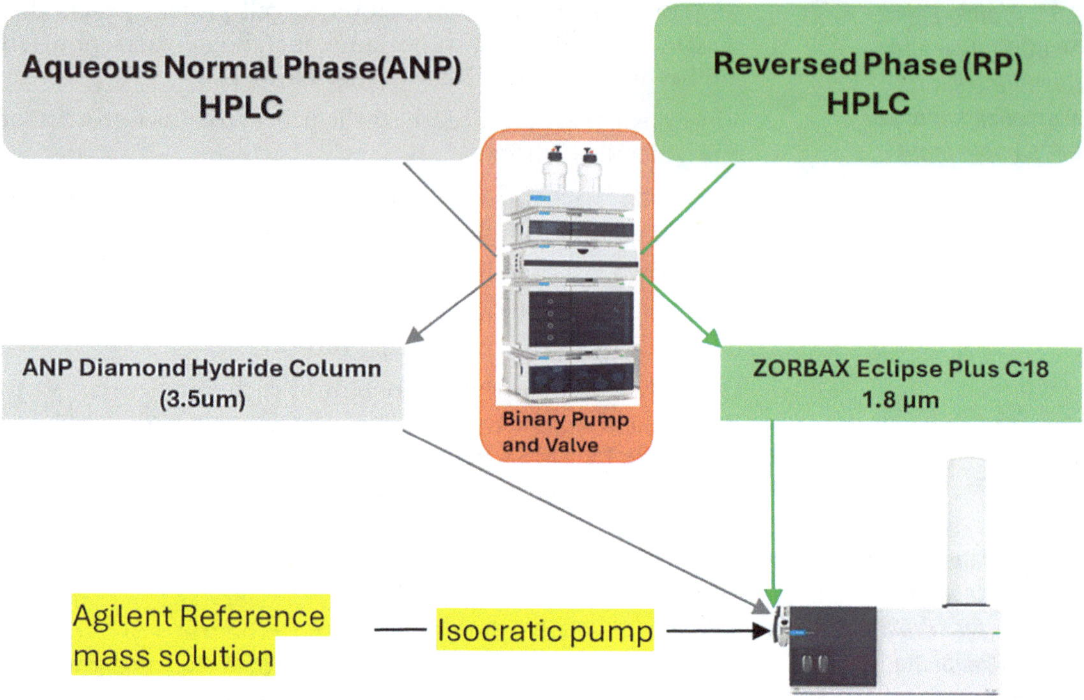

Fig. 1 LC/MS-based strategy for broad untargeted metabolite profiling (see Text for details)

isocratic pump may be used for continual delivery of reference masses to the MS system for accurate mass calibration.

Many commercial LC/MS instrument vendors provide reliable platforms for accurate mass metabolite profiling. As described previously [3], for untargeted metabolite profiling, our lab routinely uses an electrospray LC/MS platform comprised of an Agilent Model 1290 Infinity II liquid chromatography system coupled to an Agilent 6550 iFunnel time-of-flight MS analyzer (*see* **Notes 8–12**).

3.6 Metabolomic Data Analysis and Interpretation

To ascertain molecular identities for apparent differentially expressed features, LC/MS data may be searched against the publicly available and downloadable HMDB database (https://hmdb.ca/downloads) or against an alternative in-house annotated personal metabolite database that one can created from pure reference molecules using MassHunter PCDL manager 10.0 (Agilent Technologies) based on monoisotopic neutral masses (<5 ppm mass accuracy) and chromatographic retention times. Notably, a molecular formula generator algorithm in MPP is used to generate and score empirical molecular formulae, based on a weighted consideration of monoisotopic mass accuracy, isotope abundance ratios, and spacing between isotope peaks. A tentative compound ID is assigned by MPP when the PCDL database and MFG scores concur for a given candidate molecule. Tentatively assigned molecular

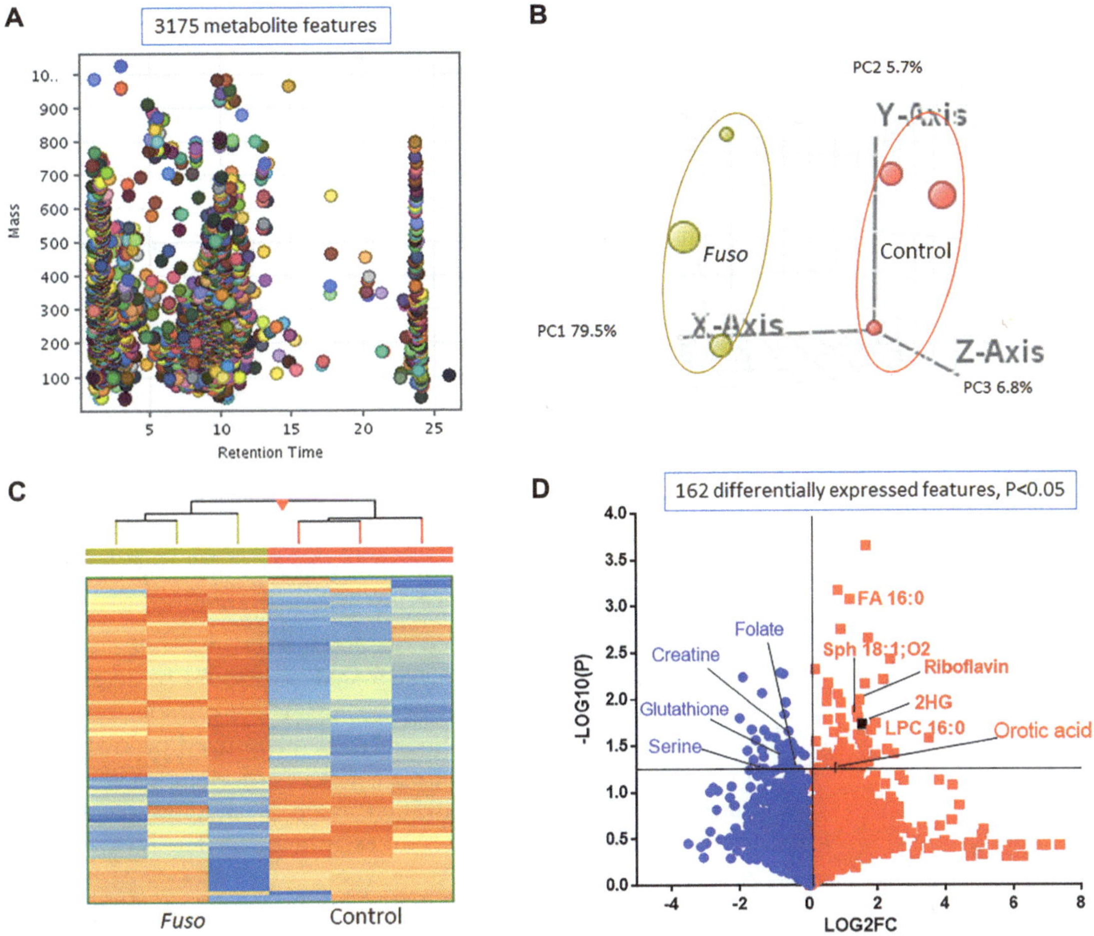

Fig. 2 Untargeted metabolite profiling identifies differentially expressed metabolites in *F. nucleatum* exposed HCT116 cell spheroids in culture. As proof of principle, HCT116 colonocytes were plated at ~150 organoids per well and grown apical side out for 72 h in 24-well ultralow attachment plates. Organoids were then cultured for an additional 24 h in antibiotic-free media, followed by administration of *F. nucleatum* at an MOI of 1:10 and incubated for an additional 72. (**a**) A total of 3175 aligned molecular features detected as negative ions were quantified in each of 6 samples by untargeted molecular feature extraction. (**b**) Principal component analysis showing distinct group separations. (**c**) Heat map from hierarchical clustering analysis, showing differentially expressed metabolites between the treatment groups. The relative level of expression for each metabolite is represented by a color-coded heat map, where red represents an expression level greater than the mean and blue represents expression below the mean. (**d**) Shown is a volcano plot depicting significant changes in metabolite abundances considering fold-change on the x-axis and statistical significance on the y-axis

identities can also be independently verified based on a match of LC retention times and/or MS/MS fragmentation spectra that are obtained from the analysis of a pure molecular reference standard. As a proof of principle, Fig. 2 depicts the results of an exemplary untargeted metabolite profiling study comparing 3D spheroids derived from HCT116 human CRC cells that were either exposed to *F. nucleatum* or vehicle (control) (*see* **Notes 13** and **14**).

4 Notes

1. Notably excess salt interferes with feature detection by mass spectrometry.
2. Knowledge of the cell protein content is necessary to normalize LC/MS injection volumes of cell extracts, to allow for meaningful ratiometric quantification of treatment-evoked changes in diverse cellular metabolites.
3. To minimize sample degradation and metabolite losses, do not leave the dry-down mixture overnight in the SpeedVac.
4. Stainless-steel beads can remain in place during the cell protein solubilization. The volume of NaOH needed for complete protein solubilization may need to be increased in the case of an overly large cell pellet.
5. Occasional vortexing (every 5 min or so) may be necessary to fully solubilize tissue protein with periodic opening of Eppendorf tube caps for release of any gaseous buildup.
6. The optimal volume of acetonitrile:H_20 reconstitution of dried metabolite extracts for LC/MS analysis should be selected to give a desired final equivalent of 6 μg/μL pf protein relative to the original cell harvest. Notably, all recognized perturbations in metabolite ratios for *F. nucleatum*-exposed cells/tumor vs. control are quantified based on metabolites derived from equivalent extract volumes, normalized to cell protein content.
7. Avoid transferring supernatant residues from the very bottom of the Eppendorf centrifuge tube as they may contain insoluble protein or other cell debris. We seek to inject a standardized 5 μL of sample volume for LC/MS analysis in both positive- and negative-ion MS analytical modes. As a rule of thumb, from a 50-μL total resuspension volume, it is best to leave ~10 μL in the tube—utilizing no more than 40 μL of the supernatant for LC/MS-based metabolite profiling analysis.
8. Robust hydrophilic/polar metabolites separations can be performed using columns from many vendors. Our lab utilizes Aqueous Neutral Phase (ANP) chromatography performed on a 100 × 2.1-mm Diamond Hydride column with 3.5-μm particle size (Microsolv). Alternatively, numerous other hydrophilic chromatographic matrices can substitute for ANP chromatography.
9. For ANP chromatography, mobile phases may consist of (A) 50% isopropanol, containing 0.025% acetic acid and (B) 90% acetonitrile containing 5-mM ammonium acetate. To eliminate interference by metal ions on chromatographic peak integrity and electrospray ionization, EDTA should be added

to the mobile phase at a final concentration of 5 μM. A typical sample Injection volume is 5 μL with a mobile phase flow rate of 0.4 mL/min. An effective mobile phase gradient for metabolite resolution is as follows: 0–1.0 min, 99% B; 1.0–15.0 min, to 20% B; 15.0–29.0 min, 0% B; and 29.1–37 min, 99% B.

10. For analysis of hydrophobic and relatively nonpolar metabolites separations, reverse phase (RP) chromatography may be effectively employed. While many commercially available RP columns can serve well, my lab routinely uses an Agilent ZORBAX Eclipse Plus C18, 100 × 2.1 mm, reversed phase column with 1.8-μm particle size.
11. If using RP chromatography with an Agilent Zorbax column, mobile phases may comprise: (A) 10 mM of ammonium formate with 5 μM of Agilent deactivator additive in 5:3:2 water: acetonitrile:2-propanol and (B) 10 mM of ammonium formate in 1:9:90 water:acetonitrile:2-propanol. Column temperature is preferably set at 55 °C with an autosampler temperature at 20 °C. Sample injection volume is 4 μL; the mobile phase flow rate is 0.4 mL/min. The following mobile phase gradient is recommended: 0 min, 15% B; 0–2.5 min, to 50% B; 2.5–2.6 min, to 57%, 2.6–9 min, to 70% B; 9–9.1 min, to 93% B; 9.1–11.1 min, to 96%; 11.1–15 min, 100% B; and 15–20 min, 15% B.
12. In our hands, acquired raw data are analyzed using MassHunter Profinder 10.0 and MassProfiler Professional (MPP) 15.1 software (Agilent technologies). Student t-tests ($p < 0.05$) are performed to identify metabolites that exhibit significantly different abundances between treatment groups (i.e., *F. nucleatum*-exposed vs. control). It would be appropriate to establish a preset statistical fold-change and p-value threshold for recognition of molecules whose abundance are specifically altered in response to a given cell treatment, e.g., colonocyte infection with *F. nucleatum*.
13. In untargeted metabolite profiling studies, recognition of one or more significant differentially expressed metabolites between treatment groups (e.g., control vs. *F. nucleatum*-exposed colonocytes) is based upon analysis of variance (ANOVA), by comparing the within-group vs. between-group variation in MS ion count abundances for all observed molecular features/metabolites. To allow for a robust ANOVA analysis, a minimum group of $n = 5$ biological replicate samples per group is highly recommended.
14. Identification of significant between-group differentially expressed metabolites from a single untargeted profiling experiment necessitates validation by repeating the finding(s) upon analysis of an independent sample cohort. Alternatively, validation of untargeted profiling findings can be provided by results from independent orthologous confirmatory studies.

Conflict of Interest

The authors have no conflicts of interest to declare that are relevant to the content of this chapter.

References

1. Sameni F, Elkhichi PA, Dadashi A, Sadeghi M, Goudarzi M, Eshkalak MP, Dadashi M (2025) Global prevalence of *Fusobacterium nucleatum* and *Bacteroides fragilis* in patients with colorectal cancer: an overview of case reports/case series and meta-analysis of prevalence studies. BMC Gastroenterol 25(1):71. https://doi.org/10.1186/s12876-025-03664-x
2. Galasso L, Termite F, Mignini I, Esposto G, Borriello R, Vitale F, Nicoletti A, Paratore M, Ainora ME, Gasbarrini A, Zocco MA (2025) Unraveling the role of *Fusobacterium nucleatum* in colorectal cancer: molecular mechanisms and pathogenic insights. Cancers (Basel) 17(3). https://doi.org/10.3390/cancers17030368
3. Chen Q, Kirk K, Shurubor YI, Zhao D, Arreguin AJ, Shahi I, Valsecchi F, Primiano G, Calder EL, Carelli V, Denton TT, Beal MF, Gross SS, Manfredi G, D'Aurelio M (2018) Rewiring of glutamine metabolism is a bioenergetic adaptation of human cells with mitochondrial DNA mutations. Cell Metab 27(5):1007–1025 e5. https://doi.org/10.1016/j.cmet.2018.03.002

Chapter 17

A Mouse Model for *Fusobacterium nucleatum*-Mediated Experimental Periodontitis

Sharon Kim, Chungyu Chang, Hung Ton-That, No-Hee Park, and Reuben H. Kim

Abstract

Fusobacterium nucleatum is a Gram-negative, anaerobic bacterium that naturally colonizes the human oral cavity and is recognized as an etiological agent in both local and systemic diseases, including periodontitis, preterm birth, and colorectal cancer. To advance research into the systemic impact of oral-origin *F. nucleatum*-associated pathologies, it is crucial to optimize the use of animal models. In this study, we present a detailed protocol for establishing a mouse model of experimental periodontitis through oral inoculation of bacteria, including *F. nucleatum*. This model provides a valuable platform for investigating the pathogenic mechanisms and systemic consequences of oral bacterial infections.

Key words *Fusobacterium nucleatum*, Mouse model, Ligature, Alveolar bone loss

1 Introduction

Fusobacterium nucleatum is a Gram-negative, anaerobic bacterium that resides as a commensal organism in the human oral cavity. It has been implicated in various oral diseases, most notably periodontitis [1]. Emerging evidence suggests that *F. nucleatum* originating from the oral cavity may also contribute to systemic pathologies, including preterm birth and colorectal cancer (CRC) [2, 3]. To elucidate the mechanistic link between oral colonization by *F. nucleatum* and its potential role in distant systemic diseases, it is essential to employ animal models that enable causal investigation through biochemical and molecular analyses, such as histology.

Several rodent models have been developed to mimic human periodontitis, including those induced by high-carbohydrate diets or by mechanical inflammation via ligature placement [4–6]. However, these models are limited in their ability to study human-derived bacteria, as rodents do not naturally harbor human-specific microbial communities [7].

Hung Ton-That (ed.), *Fusobacterium nucleatum: Methods and Protocols*, Methods in Molecular Biology, vol. 3055, https://doi.org/10.1007/978-1-0716-5388-3_17,

To address this limitation, experimental models have been established using direct oral inoculation of human bacterial strains such as *Porphyromonas gingivalis* and *Aggregatibacter actinomycetemcomitans* [8, 9]. While oral application of live bacteria is a common approach, its impact on alveolar bone loss is often modest, even with frequent high-concentration inoculations [10, 11]. This underscores the need for improved methods to establish and maintain bacterial colonization in the murine oral cavity.

Recent studies have demonstrated that transient ligature placement to create a niche around the tooth, followed by local inoculation of live bacteria, effectively induces alveolar bone loss in mouse models of bacteria-driven periodontitis [12]. Building on this approach, we present a detailed protocol for establishing a *F. nucleatum*-mediated periodontitis model in mice using ligature-assisted bacterial inoculation. This method not only facilitates the study of *F. nucleatum*-induced periodontitis but is also adaptable for investigating other oral pathogens.

2 Materials

2.1 Mouse Preparation

1. C57BL/6 J mice from available vendors, e.g., Jackson Laboratory (*see* **Note 1**).
2. 20 mg of kanamycin.
3. 20 mg of ampicillin.
4. Drinking water.

2.2 Pre-op Mouse Anesthesia

1. 100 mg/mL of ketamine.
2. 20 mg/mL of xylazine.
3. Saline.
4. 1-mL insulin syringes.
5. Eye lubricant ointment.

2.3 Ligature Placement

1. Dental loupe (3.4×).
2. 6-0 silk suture thread.
3. Two fine Science Tools Round Handled Needle Holders 12075-12 (*see* **Note 2**).
4. Fine scissors.
5. Eye lubricant ointment.

2.4 Post-op Mouse Management

1. Carprofen.
2. Heat pads.

2.5 Pre-op Mouse Anesthesia and F. nucleatum Inoculation

1. 100 mg/mL of ketamine.
2. 20 mg/mL of xylazine.
3. *F. nucleatum* ATCC 23726.
4. TSPA: Tryptic Soy Broth (TSB) supplemented with 1% Bacto™ peptone and 1.5% agar.
5. TSPC: Tryptic Soy peptone liquid medium containing 0.05% L-cysteine.
6. An anaerobic chamber.
7. Phosphate-buffered saline (PBS).
8. Sterile 1% methyl cellulose solution.
9. Topical eye lubricant ointment.
10. Heat pads.
11. Isoflurane (for euthanasia, tissue collection).
12. Scissors.

2.6 Mouse Euthanasia

1. Induction chamber.
2. Isoflurane (for euthanasia, tissue collection).
3. Scissors.

2.7 Tissue Procurement

1. PBS, pH 7.4.
2. 4% paraformaldehyde (PFA) in PBS, pH 7.4.
3. TRIzol reagent.
4. 70% ethanol solution.
5. 1-mL 29G insulin syringes.
6. Single-edge razor blade.
7. Size 15 surgical blade and blade holder.
8. Conical tubes (50 mL).

2.8 Micro Computed Tomography (μCT) Scanning and Analysis

1. SkyScan1275, Bruker- μCT Systems.
2. Voxel (size 20 μm^3).
3. Nrecon program, Bruker- μCT Systems.
4. CTAn program, Bruker- μCT Systems.
5. CTVol program, Bruker- μCT Systems.
6. 70% ethanol solution.

2.9 Tissue Preparation (Fixation and Formalin-Fixed Paraffin-Embedded Sectioning)

1. 5% EDTA in PBS, pH 7.4.
2. Microtome (e.g., Thermo Fisher Scientific, Inc).
3. 4% paraformaldehyde in PBS, pH 7.4.

2.10 Tissue Staining

1. Hematoxylin and Eosin Stain Kit.
2. Xylene.
3. 100% ethanol solution.
4. 90% ethanol solution.
5. 70% ethanol solution.

3 Methods

3.1 Mouse Preparation

1. Order C57BL/6 J mice and quarantine for 1 week before the experiment starts.
2. Four days before the ligature placement, mix ampicillin (0.5 mg/mL) and kanamycin (0.5 mg/mL) into drinking water. The water with antibiotics is left in the cage for the mice to freely drink.
3. Replace the antibiotic water daily for four consecutive days (*see* Fig. 1).

3.2 Pre-op Mouse Anesthesia

1. On the day of ligature placement (Day 0, *see* Fig. 1), prepare the anesthesia cocktail by mixing ketamine (100 mg/mL) and xylazine (20 mg/mL), and then dilute with saline (e.g., ketamine 16 μL + xylazine 4 μL + saline 80 μL).
2. Administer 100 μL of the anesthetic cocktail via intraperitoneal injection using a 1-mL insulin syringe.
3. Apply topical eye lubricant ointment to prevent corneal dryness.
4. Confirm adequate anesthesia by performing a positive toe pinch reflex test. Once verified, proceed with ligature placement on the mice.

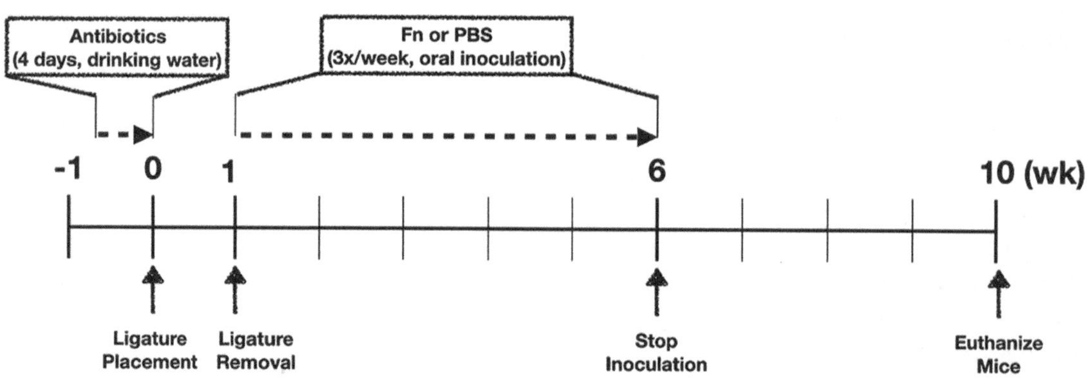

Fig. 1 Schematic diagram of the research design

3.3 Ligature Placement (see Note 3)

1. Place a mouth opener between the maxillary and mandibular incisors to keep the mouth open.
2. Position the mouse so that the head is facing upward.
3. Secure the mouth opener with a tape.
4. Tightly grasp about 10 cm of 6-0 silk suture using two round handled needle holders.
5. Floss the suture between the second and third molar.
6. Pull the suture buccally so the buccal end is longer than the palatal end of the suture.
7. Floss the longer end of the suture between the first and second molar.
8. Pull the suture palatally to tighten the suture around the second molar.
9. Re-floss the suture between the second and third molar. Tighten the suture by pulling the suture buccally.
10. Re-floss the suture between the second and first molar. Tighten the suture by pulling the suture palatally.
11. Tie the suture by knotting three times in an alternative way.
12. Use micro-scissor, leaving about 1.5 mm of suture from the knot.

3.4 Post-op Mouse Management

1. After ligature placement, administer Carprofen subcutaneously (5–6 mg/kg).
2. Place the mice on a heat pad until they recover to prevent hypothermia.
3. Monitor the mice continuously until ambulatory, at least 1-h post-op.

3.5 Pre-op Mouse Anesthesia and F. nucleatum Inoculation

1. At week 1 after ligature placement, prepare the ketamine/xylazine cocktail, and anesthetize the mice as described in the Subheading 3.2 (*see* **Note 4**).
2. Prepare fresh *F. nucleatum* cultures as follows (*see* **Note 5**):
   (a) Culture *F. nucleatum* on TSPA plate freshly supplemented with 0.1% L-cysteine at 37 °C in an anaerobic chamber (10% H_2, 10% CO_2, and 80% N_2).
   (b) Inoculate a fresh colony from TSPA plate in TSPC, and grow for overnight until OD_{600} at ~0.5 (5 x10^8 CFU/mL).
   (c) Concentrate *F. nucleatum* by centrifugation at 6000 rpm for 10 min.

(d) Wash the pellet once with PBS, and suspend in sterile 1% methyl cellulose solution to make the final concentration to 5×10^{10} CFU/mL.

3. After anesthetizing the mice, place a mouth opener between the maxillary and mandibular incisors to keep the mouth open.
4. Secure the mouth opener with tape, and remove the ligature using micro-scissor (*see* **Note 6**).
5. Use a 10-μL micropipette to collect 10 μL of freshly prepared *F. nucleatum* in methyl cellulose solution.
6. Using the end of the pipet tip, gently pull away the palatal side of the gingival tissue from the second molar, and then dispense approximately 5 μL of *F. nucleatum* into the pocket (5×10^8 CFU per gingival pocket).
7. Dispense the remaining 5 μL of *F. nucleatum* on the buccal side of the same molar.
8. For the control group, use 10 μL of PBS in methyl cellulose solution.
9. After 5 weeks, stop inoculation, and leaver for additional 4 weeks without inoculation.

3.6 Mouse Euthanasia

1. Place the mice in an induction chamber, and administer 1–5% isoflurane (pharmaceutical grade) until the mouse is recumbent.
2. Confirm the surgical plane of anesthesia by lack of response to toe pinch, and change in respiratory character and rate.
3. Perform cervical dislocation.
4. Confirm complete euthanasia by verifying the following signs: cessation of breathing, absence of pulse, lack of corneal reflex, no response to a firm toe pinch, absence of respiratory sounds and heartbeat via stethoscope, graying of mucous membranes, and onset of rigor mortis.

3.7 Tissue Procurement, Maxilla

1. Use a single-edge razor blade to horizontally section the mouth through the oral commissure. Keep sawing through until you cut the TMJ and separate the maxilla and mandible.
2. Separate the maxilla from the mouse carcass by removing the skin, fur, and muscles.
3. Using a size 15 surgical blade, carefully separate the vertical half of the palatal tissue. Store the palatal tissue in 100 μL TRIzol reagent. Store in −20 °C until RNA extraction.
4. Once a clean maxilla with minimum skin is obtained, place it in a conical tube with 10 mL of 4% PFA. Store the maxilla in PFA at 4 °C for 24 h.

3.8 μCT Scanning and Analysis

1. After 24-h fixation of the maxilla in 4% PFA, wash the maxilla with water three times and store it in 70% ethanol at 4 °C. Keep the maxilla in 70% ethanol until μCT capturing is complete.
2. Secure the sample in a voxel (size 20 μm^3) with 15-mL conical tube attached as a sample holder. Make sure to wrap the maxilla with gauze and tightly fit it into the voxel to prevent movement. Add 70% ethanol to prevent the maxilla from drying.
3. Open SkyScan1275. For mouse bones, we use 60 kV with 1-mm aluminum filter. Set exposure to ~60%.
4. Load and secure the voxel with a maxillae sample in the scanner. Twist to tightly screw in the voxel to prevent movement.
5. Set pixel size of 10.0 μm (fits for 15 mL conical tube sample holder). Define the vertical dimension to scan.
6. Set the rotation step as 0.3°, average frames to 6, disable 360° scan, and enable random movement.
7. Begin scanning.
8. After scanning, reconstruct the data using N Recon software.
9. Perform μCT analysis using 2D and 3D image constructions using Data Viewer and CTVox software, respectively (*see* Fig. 2a for an example).

3.9 Tissue Preparation (Fixation and Formalin-Fixed Paraffin-Embedded Sectioning)

1. After μCT scanning is complete, start decalcification with 5% EDTA in PBS, pH 7.4 for 3 weeks at 4 °C. Change the decalcification solution daily.
2. Process the decalcified maxillae for paraffin embedded blocks.
3. Section the blocks using a microtome at 5 μm. For maxillae, obtain sections that have both second molars present.

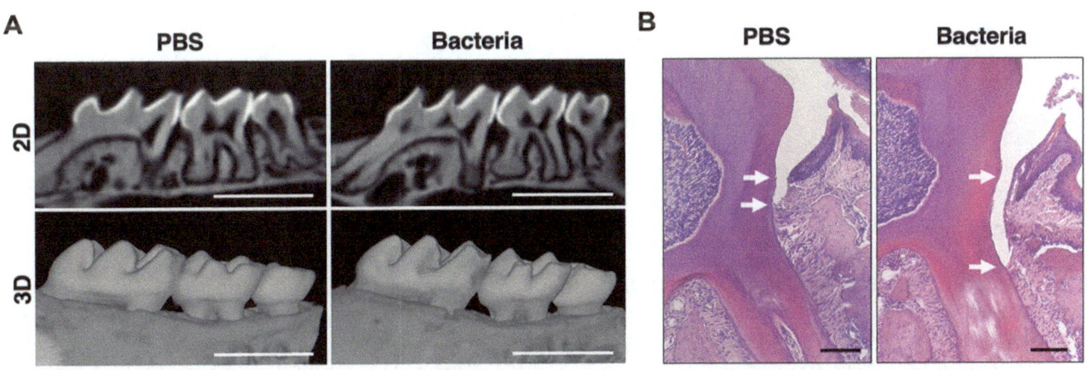

Fig. 2 Analysis of maxillae and H&E staining. (**a**) Representative two- or three-dimensional μCT images of mice maxillae. (**b**) Representative images of hematoxylin and eosin staining of the maxillary second molar and periodontal tissue. Scale bar, 100 μm. White arrows indicate the CEJ

3.10 Tissue Staining

1. Dewax the paraffin slides in xylene, and then dehydrate with 100% ethanol, 90% ethanol, and 70% ethanol solution, respectively.
2. Stain with hematoxylin and eosin (*see* Fig. 2b for an example).
3. 3.Rehydrate the slides by submerging them in 70% ethanol, 90% ethanol, 100% ethanol solution, and xylene, respectively.
4. Place the cover slip.

4 Notes

1. May use different strains of mice.
2. Using two needle holders is the key to placing the ligature atraumatically.
3. Ligature is placed for a week to create a space into which *F. nucleatum* is expected to colonize.
4. Anesthesia with ketamine/xylazine will be used every time *F. nucleatum* is inoculating.
5. *F. nucleatum* should be prepare freshly every day right before inoculation.
6. The ligature removal is done only once right before the first inoculation.

Acknowledgments

This publication was supported by the National Institute of Dental and Craniofacial Research (NIDCR) under the award numbers DE030536 and DE033281 (to RHK) and DE017382 (to HT-T). Sharon Kim was supported by the UCLA Dentist-Scientist and Oral Health-Researcher Training Program, NIDCR Grant T90DE030860.

Conflict of Interest The authors have no conflicts of interest to declare that are relevant to the content of this chapter.

References

1. Han YW (2015) *Fusobacterium nucleatum*: a commensal-turned pathogen. Curr Opin Microbiol 23:141–147. https://doi.org/10.1016/j.mib.2014.11.013
2. Han YW, Redline RW, Li M, Yin L, Hill GB, McCormick TS (2004) *Fusobacterium nucleatum* induces premature and term stillbirths in pregnant mice: implication of oral bacteria in preterm birth. Infect Immun 72(4): 2272–2279
3. Kostic AD, Chun E, Robertson L, Glickman JN, Gallini CA, Michaud M, Clancy TE, Chung DC, Lochhead P, Hold GL, El-Omar EM, Brenner D, Fuchs CS, Meyerson M,

Garrett WS (2013) *Fusobacterium nucleatum* potentiates intestinal tumorigenesis and modulates the tumor-immune microenvironment. Cell Host Microbe 14(2):207–215. https://doi.org/10.1016/j.chom.2013.07.007

4. Gupta OP, Shaw JH (1956) [Periodontal disease in the rice rat. I. Anatomic and histopathologic findings]. Oral Surg Oral Med Oral Pathol 9(6):592–603. https://doi.org/10.1016/0030-4220(56)90319-x
5. Marchesan J, Girnary MS, Jing L, Miao MZ, Zhang S, Sun L, Morelli T, Schoenfisch MH, Inohara N, Offenbacher S, Jiao Y (2018) An experimental murine model to study periodontitis. Nat Protoc 13(10):2247–2267. https://doi.org/10.1038/s41596-018-0035-4
6. Abe T, Hajishengallis G (2013) Optimization of the ligature-induced periodontitis model in mice. J Immunol Methods 394(1–2):49–54. https://doi.org/10.1016/j.jim.2013.05.002
7. Hayashi C, Gudino CV, Gibson FC 3rd, Genco CA (2010) Review: pathogen-induced inflammation at sites distant from oral infection: bacterial persistence and induction of cell-specific innate immune inflammatory pathways. Mol Oral Microbiol 25(5):305–316. https://doi.org/10.1111/j.2041-1014.2010.00582.x
8. Graves DT, Kang J, Andriankaja O, Wada K, Rossa C Jr (2012) Animal models to study host-bacteria interactions involved in periodontitis. Front Oral Biol 15:117–132. https://doi.org/10.1159/000329675
9. Oz HS, Puleo DA (2011) Animal models for periodontal disease. J Biomed Biotechnol 2011:754857. https://doi.org/10.1155/2011/754857
10. Gibson FC 3rd, Hong C, Chou HH, Yumoto H, Chen J, Lien E, Wong J, Genco CA (2004) Innate immune recognition of invasive bacteria accelerates atherosclerosis in apolipoprotein E-deficient mice. Circulation 109(22):2801–2806. https://doi.org/10.1161/01.CIR.0000129769.17895.F0
11. Lalla E, Lamster IB, Hofmann MA, Bucciarelli L, Jerud AP, Tucker S, Lu Y, Papapanou PN, Schmidt AM (2003) Oral infection with a periodontal pathogen accelerates early atherosclerosis in apolipoprotein E-null mice. Arterioscler Thromb Vasc Biol 23(8):1405–1411. https://doi.org/10.1161/01.ATV.0000082462.26258.FE
12. Kim S, Bando Y, Chang C, Kwon J, Tarverti B, Kim D, Lee SH, Ton-That H, Kim R, Nara PL, Park NH (2022) Topical application of *Porphyromonas gingivalis* into the gingival pocket in mice leads to chronicactive infection, periodontitis and systemic inflammation. Int J Mol Med 50(2). https://doi.org/10.3892/ijmm.2022.5159

Chapter 18

A Murine Floor-of-Mouth Model for Investigation of *Fusobacterium nucleatum*-Associated Oral Squamous Cell Carcinoma

Pachiyappan Kamarajan, Allan Radaic, and Yvonne L. Hernandez-Kapila

Abstract

Oral squamous cell carcinoma (OSCC) is the most prevalent malignancy of the oral cavity and a major subtype of head and neck squamous cell carcinoma (HNSCC). While established risk factors such as tobacco use, alcohol consumption, and human papillomavirus (HPV) infection contribute to OSCC, they do not fully account for its incidence or biological behavior. Emerging evidence suggests that oral pathogens, particularly those associated with periodontal disease, may contribute to carcinogenesis. Among these, *Fusobacterium nucleatum* has garnered attention for its pro-inflammatory and immunomodulatory properties. Here, we present a step-by-step protocol to evaluate the contribution of *F. nucleatum* to OSCC development using a murine floor-of-mouth tumor model. This protocol enables the assessment of microbial influences on oral tumor initiation and progression in vivo.

Key words Head and neck squamous cell carcinoma, Oral squamous cell carcinoma, *Fusobacterium nucleatum*, Orthotopic floor-of-mouth murine model, UM-SCC-14A

1 Introduction

Head and neck squamous cell carcinoma (HNSCC) ranks among the most common cancers globally, with oral squamous cell carcinoma (OSCC) accounting for approximately 90% of cases within the oral cavity [1–3]. Despite advances in diagnosis and therapy, OSCC continues to present with high morbidity and a stagnant 5-year survival rate [4, 5]. Traditional risk factors such as tobacco use, alcohol intake, and HPV infection have long been implicated [6]; however, they do not sufficiently explain all cases or the aggressive behavior observed in many patients.

Recent studies suggest a role for oral microbial dysbiosis, particularly in the context of periodontal disease, in modulating cancer risk and progression [7–12]. Periodontitis is a chronic inflammatory condition driven by a shift in the oral microbiome, often

Hung Ton-That (ed.), *Fusobacterium nucleatum: Methods and Protocols*, Methods in Molecular Biology, vol. 3055,
https://doi.org/10.1007/978-1-0716-5388-3_18,

involving well-characterized pathogenic taxa such as *Treponema denticola*, *Porphyromonas gingivalis*, and *Tannerella forsythia*—collectively known as the "red complex" [13–15]. Emerging pathogens, most notably *Fusobacterium nucleatum*, have been linked to both periodontal pathology and systemic diseases, such as colorectal cancer [16, 17]. Furthermore, we demonstrated that *F. nucleatum* promotes oral tumorigenesis in vivo [18].

F. nucleatum, a common oral bacterium, is a key bridging organism in oral biofilm development, facilitating microbial colonization and translocation across epithelial barriers [19–21]. Its enrichment in both the oral cavity and tumor microenvironments suggests a potential mechanistic link between microbial colonization and oncogenic transformation [22]. Specifically, *F. nucleatum* has been implicated in colorectal cancer progression [16, 17, 23], with identical strains detected in both colorectal tumors and the oral cavity of affected individuals [24]. *F. nucleatum* is also implicated in the development and progression of oral cancer, particularly OSCC [4, 18, 22, 25–30]. Given the importance of *F. nucleatum* in oral cancer biology, we developed an *F. nucleatum*-associated floor-of-mouth OSCC mouse model, which we describe in detail in this protocol report. This method allows for controlled microbial exposure and tumor induction, enabling mechanistic studies of microbial-host interactions in oral cancer development.

2 Materials

2.1 Culture of Oral Cancer Cells

1. UM-SCC-14A oral squamous carcinoma cells (*see* **Note 1**) [31].
2. Supplemented Dulbecco's Modified Eagle Medium (DMEM): To DMEM (Gibco) media, add 10% fetal bovine serum (FBS; Gibco), 1% penicillin (10,000 U/mL), and 1% streptomycin (10,000 μg/mL).
3. Phosphate-buffered saline (PBS; 1X, pH 7.4).
4. 100-mm Petri dishes.

2.2 Culture of F. nucleatum

1. *F. nucleatum* (ATCC 25586).
2. Brain heart infusion (BHI) broth: 5 g/L of beef heart (infusion from 250 g), 12.5 g/L of calf brains (infusion from 200 g), 2.5 g/L of disodium hydrogen phosphate, 2 g/L of D(+)-glucose, 10 g/L of peptone, and 5 g/L of sodium chloride.
3. Supplemented BHI media: 5 μg/mL of hemin and 10 μg/mL of vitamin K in BHI broth.

2.3 Preparation of UMSCC-14A Cells with F. nucleatum

1. To DMEM (Gibco) media, add 10% fetal bovine serum (FBS; Gibco), 1% penicillin (10,000 U/mL), and 1% streptomycin (10,000 μg/mL).
2. 0.05% Trypsin-EDTA.
3. Biosafety cabinet.
4. Inverted microscope (Nikon Eclipse 50i).
5. Refrigerated centrifuge.
6. Tissue culture incubator (37 °C, 5% CO_2).
7. Spectrophotometer.
8. Anaerobic chamber (90% N_2, 5% H_2, 5% CO_2) at 37 °C.

2.4 Injection of Tumor Cells

1. Athymic nude mice (NCr-nu/nu, NCI, Frederick; Strain #553) (*see* **Note 2**).
2. Growth factor-reduced Matrigel (Corning) (*see* **Note 3**).
3. Tumor injection tools: 25-gauge needle, 1-mL syringe (*see* **Note 4**).
4. Anesthetic agents: 5.0 mL (100 mg/kg) of ketamine (*see* **Note 5**), xylazine.
5. Animal facility.
6. Calipers for tumor measurement.
7. CO_2 euthanasia chamber.

3 Methods

3.1 Culture of Oral Cancer Cells

1. Thaw and culture UM-SCC14A cells in supplemented DMEM.
2. Maintain cells at 37 °C in a humidified incubator with 5% CO_2.
3. Subculture cells using 0.05% Trypsin-EDTA until they are 70–80% confluent.

3.2 Culture of F. nucleatum

1. Thaw *F. nucleatum* (ATCC 25586) and inoculate into supplemented BHI broth.
2. Incubate cultures anaerobically at 37 °C using a chamber flushed with 90% N_2, 5% H_2, and 5% CO_2.
3. Grow for 24–48 h until mid-log phase.
4. Confirm culture purity via Gram staining and light microscopy.
5. Adjust bacterial concentration to OD_{600} = 0.1 (~1×10^8 CFU/mL) using sterile PBS.

3.3 Challenge/Inoculate UM-SCC-14A Cells with F. nucleatum

1. Plate UM-SCC-14A cells in 100-mm dishes, and allow them to grow until they are 70–80% confluent.
2. Centrifuge *F. nucleatum* at 3000× *g* for 10 min at 4 °C.
3. Wash the bacterial pellet twice with sterile PBS, and resuspend in serum- and antibiotic-free DMEM.
4. Add bacteria to the cells at a multiplicity of infection (MOI) of 50.
5. Incubate for 2 h at 37 °C in a humidified incubator with 5% CO_2.
6. Wash cells three times with sterile PBS to remove non-adherent bacteria.

3.4 Injection of Tumor Cells

1. Trypsinize *F. nucleatum*-infected UM-SCC-14A cells with 0.05% trypsin-EDTA.
2. Prepare the tumor cells as a single-cell suspension.
3. Resuspend cells in DMEM and mix 1:1 with growth factor-reduced Matrigel on ice to a final concentration of 1×10^6 cells/50 μL (*see* **Note 6**).
4. Anesthetize 6-week-old athymic nude mice (NCr-nu/nu) with ketamine (100 mg/kg) and xylazine (10 mg/kg) via intraperitoneal injection (*see* **Note 7**).
5. Inject 50 μL of the cell-Matrigel suspension submucosally into the floor of the mouth using a 25-gauge needle. The mouse should be held firmly by scruffing its neck to secure it gently. The needle should be introduced at a 60-degree angle to a depth of 2–3 mm, ensuring the tongue is avoided during the process (*see* Fig. 1a) (*see* **Note 8**).
6. Monitor mice daily for tumor formation and progression (*see* **Note 9**).
7. Weigh the mice each week to track any weight gain or loss.
8. Upon tumor formation (typically by week 2), randomize mice into experimental groups.
9. Measure tumor dimensions using calipers and calculate volume using the formula:

   Volume = a × a/b × 2, where a is the shortest and b is the longest dimension.
10. At the study endpoint, euthanize mice by CO_2 overdose followed by cervical dislocation (*see* **Note 10**).
11. Excise tumors (*see* Fig. 1b), rinse in PBS, and fix in 10% neutral-buffered formalin overnight at 4 °C.
12. Embed fixed tissues in paraffin, section at 5 μm, and perform staining for histological analysis.

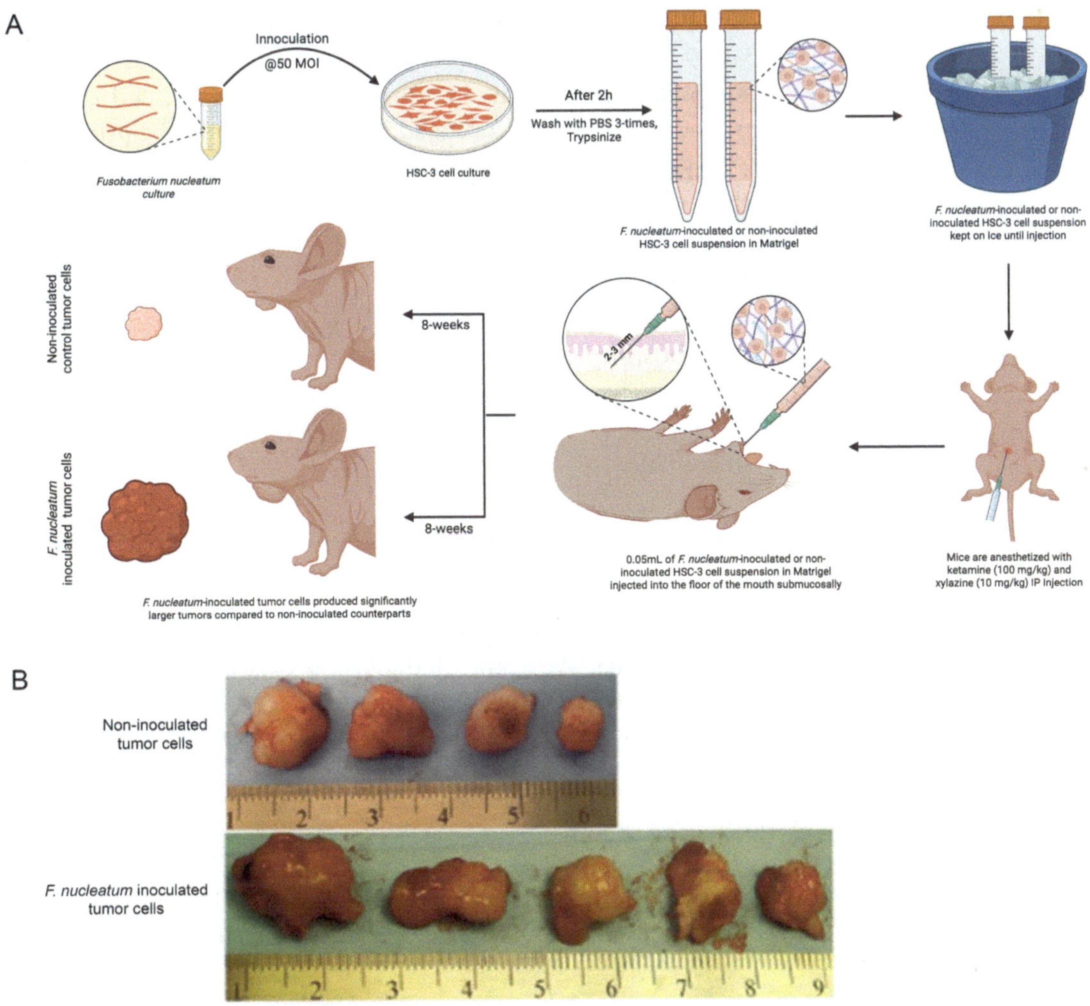

Fig. 1 (**a**) The schematic diagram illustrates the protocol for *Fusobacterium nucleatum*-associated OSCC floor-of-mouth mouse model. (**b**) *F. nucleatum* promote tumorigenesis. Images show the dissected tumors obtained from mice injected with UM-SCC-14A cells challenged with a control medium or media containing *F. nucleatum* (50 MOI). (Kamarajan et al. [18])

4 Notes

1. Select the correct cell line, and determine the required concentration for implantation according to your experimental goals.
2. Obtaining animal use approval for animal experimentation prior to the start of any project is required.
3. Sterilize the surgical site and all tools and instruments used to mitigate infection risks.
4. Ensure that all reagents and supplies used on animals are medical/veterinary grade.

5. Ketamine is a schedule III controlled substance can be procured from the institutional pharmacy.
6. Keep all materials, including Matrigel, syringes, and cell suspensions, on ice to prevent solidification during the procedure.
7. Carefully monitor the depth of anesthesia in mice, ensuring they are fully unconscious before injecting tumor cells.
8. Use a volume of ≤50 μL for the cell injection to prevent leakage and ensure accurate placement.
9. Observe the mice closely for any signs of distress or complications postimplantation.
10. Establish clear criteria for euthanasia, including significant weight loss or tumor burden, to prevent unnecessary suffering in animals.

Acknowledgments

We acknowledge support from NIH grant R01DE025225 to Yvonne L. Kapila. We acknowledge using ChatGPT (https://chat.openai.com/) for assistance in summarizing the background and methods sections from our previously published article by Kamarajan et al. (2020) Plos Pathog 16(10): e1008881. https://doi.org/10.1371/journal.ppat.100888. Figure 1a was created with Biorender.com.

Conflict of Interest The authors have no conflicts of interest to declare that are relevant to the content of this chapter.

References

1. Sung H, Ferlay J, Siegel RL, Laversanne M, Soerjomataram I, Jemal A, Bray F (2021) Global cancer statistics 2020: GLOBOCAN estimates of incidence and mortality worldwide for 36 cancers in 185 countries. CA Cancer J Clin 71(3):209–249. https://doi.org/10.3322/caac.21660
2. Zhou T, Huang W, Wang X, Zhang J, Zhou E, Tu Y, Zou J, Su K, Yi H, Yin S (2024) Global burden of head and neck cancers from 1990 to 2019. iScience 27(3):109282. https://doi.org/10.1016/j.isci.2024.109282
3. Badwelan M, Muaddi H, Ahmed A, Lee KT, Tran SD (2023) Oral squamous cell carcinoma and concomitant primary tumors, what do we know? A review of the literature. Curr Oncol 30(4):3721–3734. https://doi.org/10.3390/curroncol30040283
4. Shanmugam SB, Vijayashree Priyadharsini J, Anitha P, Smiline Girija AS, Paramasivam A (2025) Dysregulated genes in HIGK-treated F. nucleatum and their possible association with HNSCC. Mol Biol Res Commun 14(1): 59–71. https://doi.org/10.22099/mbrc.2024.50171.1982
5. Gormley M, Creaney G, Schache A, Ingarfield K, Conway DI (2022) Reviewing the epidemiology of head and neck cancer: definitions, trends and risk factors. Br Dent J 233(9):780–786. https://doi.org/10.1038/s41415-022-5166-x
6. Barsouk A, Aluru JS, Rawla P, Saginala K, Barsouk A (2023) Epidemiology, risk factors, and prevention of head and neck squamous cell carcinoma. Med Sci (Basel) 11(2). https://doi.org/10.3390/medsci11020042

7. Devaraja K, Aggarwal S (2025) Dysbiosis of oral microbiome: a key player in oral carcinogenesis? A critical review. Biomedicine 13(2). https://doi.org/10.3390/biomedicines13020448
8. Mivehchi H, Eskandari-Yaghbastlo A, Pour Bahrami P, Elhami A, Faghihinia F, Nejati ST, Kazemi KS, Nabi Afjadi M (2025) Exploring the role of oral bacteria in oral cancer: a narrative review. Discov Oncol 16(1):242. https://doi.org/10.1007/s12672-025-01998-2
9. Radaic A, Shamir ER, Jones K, Villa A, Garud NR, Tward AD, Kamarajan P, Kapila YL (2023) Specific oral microbial differences in proteobacteria and bacteroidetes are associated with distinct sites when moving from healthy mucosa to Oral dysplasia-a microbiome and gene profiling study and focused review. Microorganisms 11(9). https://doi.org/10.3390/microorganisms11092250
10. La Rosa GRM, Gattuso G, Pedulla E, Rapisarda E, Nicolosi D, Salmeri M (2020) Association of oral dysbiosis with oral cancer development. Oncol Lett 19(4):3045–3058. https://doi.org/10.3892/ol.2020.11441
11. Radaic A, Kapila YL (2021) The oralome and its dysbiosis: new insights into oral microbiome-host interactions. Comput Struct Biotechnol J 19:1335–1360. https://doi.org/10.1016/j.csbj.2021.02.010
12. Radaic A, Ganther S, Kamarajan P, Grandis J, Yom SS, Kapila YL (2021) Paradigm shift in the pathogenesis and treatment of oral cancer and other cancers focused on the oralome and antimicrobial-based therapeutics. Periodontol 2000 87(1):76–93. https://doi.org/10.1111/prd.12388
13. Marsh PD (2003) Are dental diseases examples of ecological catastrophes? Microbiology (Reading) 149(Pt 2):279–294. https://doi.org/10.1099/mic.0.26082-0
14. Darveau RP (2010) Periodontitis: a polymicrobial disruption of host homeostasis. Nat Rev Microbiol 8(7):481–490. https://doi.org/10.1038/nrmicro2337
15. Mohanty R, Asopa SJ, Joseph MD, Singh B, Rajguru JP, Saidath K, Sharma U (2019) Red complex: Polymicrobial conglomerate in oral flora: a review. J Family Med Prim Care 8(11): 3480–3486. https://doi.org/10.4103/jfmpc.jfmpc_759_19
16. Bullman S, Pedamallu CS, Sicinska E, Clancy TE, Zhang X, Cai D, Neuberg D, Huang K, Guevara F, Nelson T, Chipashvili O, Hagan T, Walker M, Ramachandran A, Diosdado B, Serna G, Mulet N, Landolfi S, Ramon YCS, Fasani R, Aguirre AJ, Ng K, Elez E, Ogino S, Tabernero J, Fuchs CS, Hahn WC, Nuciforo P, Meyerson M (2017) Analysis of Fusobacterium persistence and antibiotic response in colorectal cancer. Science 358(6369):1443–1448. https://doi.org/10.1126/science.aal5240
17. Zepeda-Rivera M, Minot SS, Bouzek H, Wu H, Blanco-Miguez A, Manghi P, Jones DS, LaCourse KD, Wu Y, McMahon EF, Park SN, Lim YK, Kempchinsky AG, Willis AD, Cotton SL, Yost SC, Sicinska E, Kook JK, Dewhirst FE, Segata N, Bullman S, Johnston CD (2024) A distinct Fusobacterium nucleatum clade dominates the colorectal cancer niche. Nature 628(8007):424–432. https://doi.org/10.1038/s41586-024-07182-w
18. Kamarajan P, Ateia I, Shin JM, Fenno JC, Le C, Zhan L, Chang A, Darveau R, Kapila YL (2020) Periodontal pathogens promote cancer aggressivity via TLR/MyD88 triggered activation of Integrin/FAK signaling that is therapeutically reversible by a probiotic bacteriocin. PLoS Pathog 16(10):e1008881. https://doi.org/10.1371/journal.ppat.1008881
19. Groeger S, Zhou Y, Ruf S, Meyle J (2022) Pathogenic mechanisms of Fusobacterium nucleatum on oral epithelial cells. Front Oral Health 3:831607. https://doi.org/10.3389/froh.2022.831607
20. Fan R, Li Z, Shi X, Wang L, Zhang X, Dong Y, Quan C (2022) Expression, purification, and characterization of the recombinant, two-component, response regulator ArlR from Fusobacterium nucleatum. Appl Biochem Biotechnol 194(5):2093–2107. https://doi.org/10.1007/s12010-021-03785-5
21. Fan Z, Tang P, Li C, Yang Q, Xu Y, Su C, Li L (2023) Fusobacterium nucleatum and its associated systemic diseases: epidemiologic studies and possible mechanisms. J Oral Microbiol 15(1):2145729. https://doi.org/10.1080/20002297.2022.2145729
22. Pignatelli P, Nuccio F, Piattelli A, Curia MC (2023) The role of Fusobacterium nucleatum in oral and colorectal carcinogenesis. Microorganisms 11(9). https://doi.org/10.3390/microorganisms11092358
23. Alon-Maimon T, Mandelboim O, Bachrach G (2022) Fusobacterium nucleatum and cancer. Periodontol 2000 89(1):166–180. https://doi.org/10.1111/prd.12426
24. Komiya Y, Shimomura Y, Higurashi T, Sugi Y, Arimoto J, Umezawa S, Uchiyama S, Matsumoto M, Nakajima A (2019) Patients with colorectal cancer have identical strains of Fusobacterium nucleatum in their colorectal cancer and oral cavity. Gut 68(7):1335–1337. https://doi.org/10.1136/gutjnl-2018-316661

25. Chang C, Geng F, Shi X, Li Y, Zhang X, Zhao X, Pan Y (2019) The prevalence rate of periodontal pathogens and its association with oral squamous cell carcinoma. Appl Microbiol Biotechnol 103(3):1393–1404. https://doi.org/10.1007/s00253-018-9475-6
26. Bronzato JD, Bomfim RA, Edwards DH, Crouch D, Hector MP, Gomes B (2020) Detection of Fusobacterium in oral and head and neck cancer samples: a systematic review and meta-analysis. Arch Oral Biol 112:104669. https://doi.org/10.1016/j.archoralbio.2020.104669
27. Neuzillet C, Marchais M, Vacher S, Hilmi M, Schnitzler A, Meseure D, Leclere R, Lecerf C, Dubot C, Jeannot E, Klijanienko J, Mariani O, Calugaru V, Hoffmann C, Lesnik M, Badois N, Borcoman E, Piaggio E, Kamal M, Le Tourneau C, Bieche I (2021) Prognostic value of intratumoral Fusobacterium nucleatum and association with immune-related gene expression in oral squamous cell carcinoma patients. Sci Rep 11(1):7870. https://doi.org/10.1038/s41598-021-86816-9
28. Harrandah AM, Chukkapalli SS, Bhattacharyya I, Progulske-Fox A, Chan EKL (2020) Fusobacteria modulate oral carcinogenesis and promote cancer progression. J Oral Microbiol 13(1):1849493. https://doi.org/10.1080/20002297.2020.1849493
29. Sun J, Tang Q, Yu S, Xie M, Zheng W, Chen G, Yin Y, Huang X, Wo K, Lei H, Zhang J, Wan Q, Chen L (2023) F. nucleatum facilitates oral squamous cell carcinoma progression via GLUT1-driven lactate production. EBioMedicine 88:104444. https://doi.org/10.1016/j.ebiom.2023.104444
30. Li Z, Liu Y, Huang X, Wang Q, Fu R, Wen X, Liu J, Zhang L (2024) F. Nucleatum enhances oral squamous cell carcinoma proliferation via E-cadherin/beta-catenin pathway. BMC Oral Health 24(1):518. https://doi.org/10.1186/s12903-024-04252-3
31. Brenner JC, Graham MP, Kumar B, Saunders LM, Kupfer R, Lyons RH, Bradford CR, Carey TE (2010) Genotyping of 73 UM-SCC head and neck squamous cell carcinoma cell lines. Head Neck 32(4):417–426. https://doi.org/10.1002/hed.21198

Chapter 19

Modeling *Fusobacterium nucleatum*-Driven Colorectal Cancer in *Apc*-Mutant Mice

Bing Han, Wei Liu, Yimin Chen, Yi-Wei Chen, Hung Ton-That, Jia Chang, and Cun-Yu Wang

Abstract

Colorectal cancer (CRC) typically initiates with mutations in the *adenomatous polyposis coli* (*Apc*) gene, followed by the influence of environmental factors. Among these, *Fusobacterium nucleatum* has emerged as an important CRC-enriched microbial species with demonstrated tumor-promoting potential. To investigate the contribution of *F. nucleatum* in the context of oncogenic mutations, physiologically relevant mouse models are essential. While $Apc^{min/+}$ mice have been employed to study *F. nucleatum*-associated CRC, prior studies have not systematically evaluated the effects of different bacterial inoculation routes and dosages on tumorigenesis. Here, we utilized a reliable and reproducible inducible *Apc*-mutant mouse model that recapitulates *F. nucleatum* colonization within the gut microenvironment, closely reflecting conditions observed in CRC patients. We provide a detailed protocol describing two distinct *F. nucleatum* administration routes and graded dosage regimens to assess their impact on CRC progression. This platform offers a robust experimental tool for investigating the microbiota-cancer axis, with broad applicability for both preclinical and mechanistic studies aimed at advancing understanding of CRC pathogenesis.

Key words *Fusobacterium nucleatum*, Colorectal cancer, *Apc*, Mouse models, FISH

1 Introduction

Colorectal cancer (CRC) represents a growing global health challenge, ranking as the third most frequently diagnosed malignancy and the second leading cause of cancer-related mortality worldwide [1]. The etiology of CRC involves a complex interplay between hereditary and environmental factors, with accumulating evidence highlighting the critical role of the gut microbiota in disease initiation and progression [2]. Among the diverse microbial species implicated in CRC, *Fusobacterium nucleatum* has emerged as one of the most consistently enriched pathogens in colorectal tumors, as demonstrated by whole-genome sequencing and metagenomic analyses [3, 4]. *F. nucleatum* is a Gram-negative, anaerobic

Hung Ton-That (ed.), *Fusobacterium nucleatum: Methods and Protocols*, Methods in Molecular Biology, vol. 3055,
https://doi.org/10.1007/978-1-0716-5388-3_19,

bacterium that inhabits the alimentary tract, from the oral cavity to the colon. Elevated *F. nucleatum* abundance has been repeatedly reported in CRC patients, correlating with advanced disease stage, metastasis, and poorer clinical outcomes, positioning *F. nucleatum* as a potential biomarker for CRC diagnosis and prognosis [5–8]. To elucidate *F. nucleatum*'s mechanistic role in colorectal tumorigenesis, multiple experimental models have been employed, including $Apc^{min/+}$ mice and xenograft systems [9–11]. However, replicating *F. nucleatum* infection in vivo remains challenging due to technical limitations in mimicking the physiological routes of bacterial entry observed in human CRC. In particular, achieving consistent and effective bacterial colonization has been problematic. Given that *F. nucleatum* can reach the intestinal mucosa either via hematogenous dissemination or by translocating through the oral-gastrointestinal tract barrier in humans [12], we established mouse models using both oral gavage and intravenous inoculation to recapitulate these distinct transmission pathways.

In this study, we describe two distinct routes of *F. nucleatum* administration, each with defined bacterial dosage regimens. We evaluate inoculation efficiency and biological effects in a genetically engineered, inducible *Apc*-mutant mouse model ($CDX2P\text{-}CreER^{T2};Apc^{15lox/15lox}$), in which tamoxifen induction initiates colon epithelium-specific *Apc* deletion, representing the earliest step in colorectal tumor development [13]. Our objective is to establish a reliable and reproducible protocol for *F. nucleatum* infection in this CRC model, enabling systematic evaluation of the pathogen's contribution to tumor progression and providing a versatile platform for microbiota-cancer research. An overview of the experimental workflow is illustrated (*see* Fig. 1).

2 Materials

2.1 Breeding, Genotyping, and Induction of Colon Tumors in Apc-Mutant Mouse

2.1.1 Mouse Strain

1. *Apc15lox* (The Jackson Laboratory, 029275).
2. $CDX2P\text{-}CreER^{T2}$ mouse (The Jackson Laboratory, 022390).

2.1.2 Genomic DNA Extraction from Mouse Tails for Genotyping

1. 1.5-mL Eppendorf tubes.
2. 50 mM of NaOH buffer.
3. 2 × PCR master mix (LAMDA Biotech).
4. Primers sequences.
   Apc15lox-F: GCTTATTTATACCCTTGGCAATC.
   Apc15lox-R: CAAAGCACACTTTGGCAGACT.
   CDX2PERT-F: CATGGTGAGGTCTGCTGATG.
   CDX2PERT-R: CATGTCCATCAGGTTCTTGC.

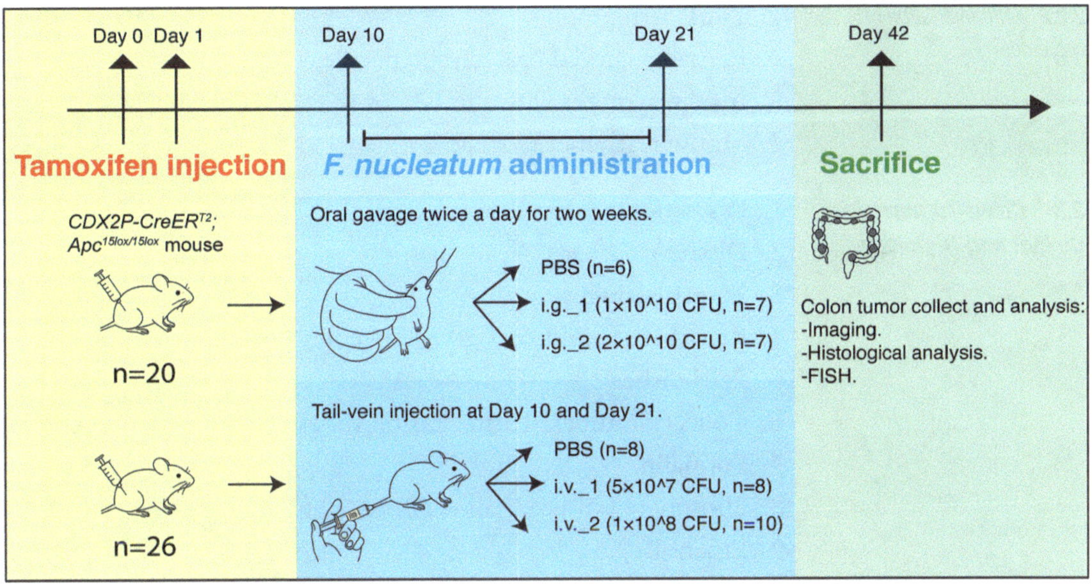

Fig. 1 Schematic of colorectal cancer induction in *Apc*-mutant mice. Tamoxifen is administered intraperitoneally to initiate tumorigenesis, followed by *F. nucleatum* delivery using two routes of administration

5. DNA loading buffer (6×).
6. 50 × TAE buffer: Dissolve 242 g of Tris base in 750 mL of H_2O, and then add 57.1 mL of glacial acid and 100 mL of 0.5 M EDTA (pH 8.0). Adjust the solution to a final volume of 1 L and pH of 8.5.
7. 2% Agarose gel with ethidium bromide (final concentration 0.25 μg/mL).
8. SYBR safe DNA gel stain.
9. 1 Kb Plus DNA ladder.

2.1.3 Tumor Induction Upon Cre Activation via Tamoxifen Administration

1. Tamoxifen (Sigma).
2. 100% EtOH.
3. Corn oil.
4. 1-mL syringe.
5. 26 G 3/8 needle.

2.2 Preparation and Administration of F. nucleatum

2.2.1 Preparation of F. nucleatum

1. *F. nucleatum* ATCC 23726.
2. TSPC: Tryptic Soy Broth (TSB) supplemented with 1% Bacto peptone and 0.25% freshly made L-cysteine.
3. Anaerobic Hungate culture tubes.
4. Phosphate-buffered saline (PBS), pH 7.4.
5. Anaerobic chamber with appropriate gas mix: 3% H_2, 5% CO_2, and 92% N_2 (*see* **Note 1**).

2.2.2 Administration of F. nucleatum

1. 1-mL syringes.
2. Reusable curved oral gavage needles (22G).
3. Insulin syringes with the 12.7 mm × 28G needle.
4. Mouse restrainer (*see* Fig. 2).

2.3 Colon Tumor Collect and Histology

1. Dissection scissors and tweezers.
2. Dissection board.
3. 20-mL syringes.
4. Reusable straight oral gavage needles (20G).
5. Petri dishes.
6. Collection tubes: 15-mL falcon tubes.
7. Formalin.
8. Phosphate buffered saline (PBS), pH 7.4.
9. Ethanol.
10. Ruler.
11. Tissue-embedding cassettes.

2.4 Fluorescence In Situ Hybridization (FISH) Analysis

1. Formalin.
2. Xylene.
3. Tissue rehydration solutions: 100% EtOH, 90% EtOH, 70% EtOH, and 50% EtOH.
4. Proteinase solution: 50 ug/mL of proteinase K in pre-warmed 50 mM of Tris-HCl.

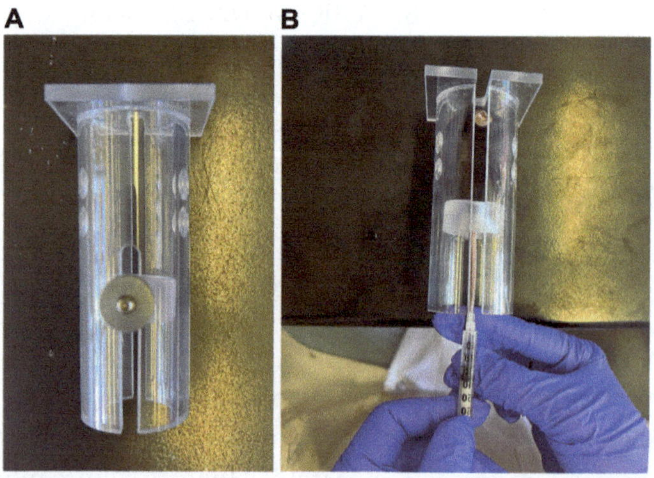

Fig. 2 Intravenous administration technique for *F. nucleatum*. (**a**) Commercially available mouse restrainer. (**b**) Injection setup showing needle placement. The tail is gently secured and aligned to provide a clear view of the vessel prior to needle insertion

5. FISH hybridization solution: 900 mM of NaCl, 20 mM of Tris-HCl (pH 7.5), 0.01% SDS, 20% v/v formamide, 2 nM of *F. nucleatum* probe, ad H_2O.
6. FISH wash buffer: 215 mM of NaCl, 20 mM of Tris-HCl (pH 7.5), 5 mM of EDTA, and H_2O.
7. Coverslips.
8. ProLong™ Diamond Antifade Mountant with DAPI (Thermo Scientific).
9. Hybridization oven.
10. Cy2-labeled *F. nucleatum* probe (FUS664): 5'-CTT GTA GTT CCG CYT ACC TC-3'.

3 Methods

3.1 Breeding, Genotyping, and Induction of Colon Tumors in Apc-Mutant Mouse

We established an inducible *Apc* knockout system in the colonic epithelium using the *Cre-LoxP* strategy. This system employs two transgenic mouse lines: one carrying *Apc* exon 16 flanked by *LoxP* sites and another expression an inducible *Cre* transgene under a colon epithelium-specific promoter. Crossbreeding these two lines generated conditional knockout mice in which *Apc* can be selectively inactivated in the colonic epithelium upon Cre induction. Homozygous deletion of the last exon of *Apc* results in rapid and efficient adenoma formation, closely resembling the phenotype observed in $Apc^{min/+}$ mice but appearing earlier and restricted to the colon. All breeding and experimental procedures were approved by the UCLA Animal Research Committee and conducted in accordance with institutional guidelines. Breeding cages were set up by housing two female mice (7–14 weeks old) with one male mouse (7–14 weeks old). Based on genotyping results, only $CDX2P\text{-}CreER^{T2};Apc^{15lox/15lox}$ compound mice (hereafter referred to as *Apc*-mutant mice) were selected for further breeding or experimental use.

3.1.1 Genomic DNA Extraction from Mouse Tails for Genotyping

1. When the mice are weaned, cut 5 μL from the tail to isolate genomic DNA for genotyping, and make a toe tattoo for identification. Follow the UCLA Animal Research Committee guidelines.
2. Put the small tailpiece in a marked 1.5-mL Eppendorf tube (*see* **Note 2**).
3. Add 300 μL of complete lysis buffer (50 mM of NaOH) to the marked tube containing a single tail biopsy.
4. Incubate the tubes 95 °C in a heat block for 1 h (*see* **Note 3**).
5. Invert the tubes several times.

6. Spin the tubes for 5 min at full speed in a centrifuge to obtain a clear supernatant.
7. Use the supernatant to perform PCR assays to check for the presence of absence Apc-LoxP and Cre using specific primers. The PCR reaction and PCR cycle conditions used for genotyping are described below.

PCR Reaction Mixture for Genotyping.

Template DNA	2 μL
Primer pair (5 μM)	2 μL
2X PCR master mix	10 μL
ddH_2O	6 μL

8. PCR cycle conditions for genotyping.
   Step 1: 94 °C for 3 min.
   Step 2: 94 °C for 30 s (denaturation).
   Step 3: 60 °C for 15 s (annealing).
   Step 4: 68 °C for 60 s (extension).
   Step 5: Step 2–5 35 times (cycling).
   Step 6: 68 °C for 5 min (final extension).
   Step 7: 4 °C until removal (end hold).
9. Analysis of PCR products: Once the PCR reaction is complete, add 5 μL 6 × DNA loading buffer to the samples and run these on a 2% agarose gel in 1 × TAE buffer. The presence or absence of the amplified PCR products will be visible under UV light. The size of the specific amplicons for *Apc-LoxP* and Cre is 380 bp (wild type is 225 bp), 200 bp, resp. (*see* **Note 4**).

3.1.2 Tumor Induction Upon Cre Activation via Tamoxifen Administration

1. Dissolve 200 mg of tamoxifen in 800 μL of 100% EtOH. Vortex well for 2 min.
2. Add 9.2 mL of corn oil. Vortex for 2–4 h (*see* **Note 5**).
3. Aliquot and store the stock solution (2 mg per 100 μL) at −20 °C (*see* **Note 6**).
4. Thaw a tamoxifen aliquot at 37 °C.
5. Vortex the tamoxifen solution for 1 min.
6. Using a 1-mL syringe fitted with a 26G, 3/8-inch needle, administer tamoxifen intraperitoneally (2 mg per 20–25-g mouse for 2 consecutive days) by injecting into the lower abdominal quadrant between the inner thigh and genital region to minimize the risk of organ puncture (*see* **Note 7**).
7. Inject slowly, and leave needle in for about 2 s to minimize back flow or solution loss.

3.2 Preparation and Administration of F. nucleatum

3.2.1 Preparation of F. nucleatum

1. Inoculate *F. nucleatum* anaerobically in TSPC medium at 37 °C overnight (12–16 h).
2. On the day of administration, dilute the overnight culture in fresh, pre-reduced TSPC to OD_{600} of 0.1. Incubate anaerobically at 37 °C for approximately 4 h or until the culture reaches $OD_{600} \approx 0.5$.
3. Harvest the bacteria by centrifugation at 8000× *g* for 5 min.
4. Resuspend the bacterial pellet in sterile PBS to the desired CFU concentrations (see below). Keep the suspensions on ice and use immediately.
5. Prepare two fusobacterial suspensions at 5×10^{10} and 1×10^{11} CFU/mL in PBS for two treatment groups, administered by both oral gavage (og) and intravenous injection (iv). Each mouse will receive 1×10^{10} and 2×10^{10} CFU, respectively, in a total volume of 200 μL (og) or 100 μL (iv).

3.2.2 Administration of F. nucleatum via Oral Gavage

1. Restrain the mouse by scuffing: Grasp the skin over the shoulders using the thumb and middle finger while stabilizing the forelimbs laterally to prevent interference with the gavage tube. Gently extend the head to align the neck and esophagus for safe tube insertion.
2. Insert the gavage needle into the diastema, and guide it gently along the upper palate until reaching the esophagus. Allow the tube to advance without resistance. In one smooth motion, deliver 200 μL of the bacterial suspension (*see* **Note 8**).
3. Remove the gavage needle gently, maintaining the same angle used during insertion.
4. Return the mouse to its cage and observe for 5–10 min for signs of distress or labored breathing. Reassess the animals 24 h after dosing.

3.2.3 Administration of F. nucleatum via Intravenous Injection

1. Restrain the mouse using an appropriately sized, commercially available restraint device. Minimize the duration of restraint, and clean the apparatus frequently to reduce pheromone-mediated stress and prevent cross-contamination between animals.
2. Using the dominant hand, stabilize the syringe near the plunger, ensuring precise control during injection. Deliver 100 μL of the bacterial suspension smoothly into the vessel without shifting the needle's position within the vein (*see* Fig. 2 and **Note 9**).
3. Withdraw the needle gently and apply light pressure to the injection site until bleeding has ceased.
4. Return the mouse to its cage and monitor to ensure that bleeding does not recur.

3.3 Colon Tumor Collect and Histology

In *Apc*-mutant mice, colonic dysplastic lesions emerge as early as 1 week after tamoxifen induction. By 5–7 weeks post-induction, these lesions progress into macroscopically visible adenomas that display characteristic histopathological features of early colon tumorigenesis. In this protocol, colonic tissues are harvested 6 weeks after tamoxifen treatment to assess *F. nucleatum* colonization and evaluate its effects on tumor development. Representative gross images and corresponding hematoxylin and eosin (H&E)-stained sections illustrating colonic morphology are shown in Fig. 3.

1. Euthanize the mouse using a rising concentration of CO_2 in accordance with UCLA Animal Research Committee guidelines.
2. Position the mouse on a dissection board, secure all four limbs, and spray the carcass with ethanol.
3. Using dissection scissors and forceps, carefully open the peritoneal cavity to expose the abdominal organs.
4. Identify the cecum and transect the colon at its junction with the cecum. Remove associated adipose tissue, and then cut at the anus to isolate an ~7–8-cm segment of colon (*see* **Note 10**).
5. Flush the colon with cold PBS to remove fecal material. Hold the tissue gently with forceps, and slowly introduce PBS into the lumen using a 20G gavage needle attached to a 20-mL

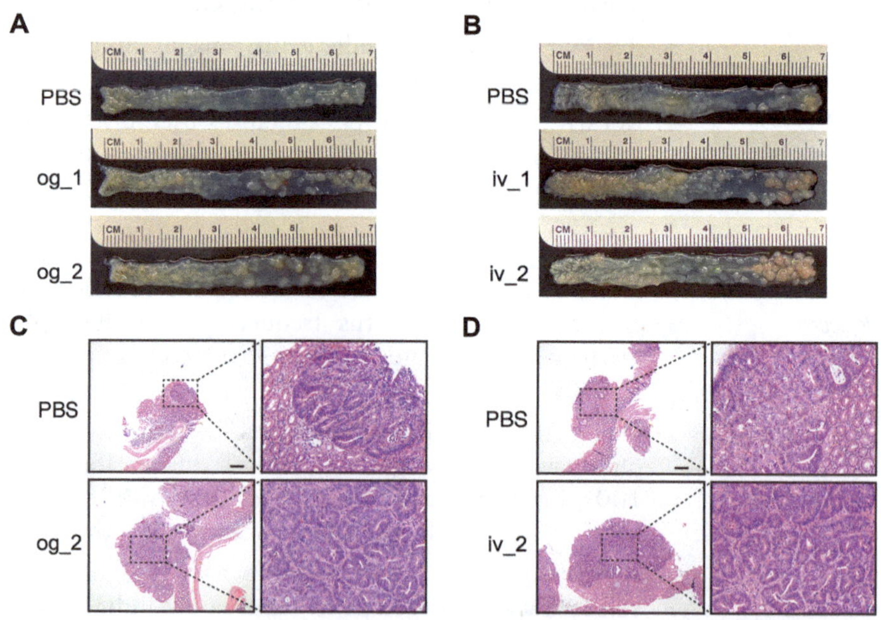

Fig. 3 Histological analysis of colon tumors following *F. nucleatum* administration. Representative images of resected colons from mice receiving *F. nucleatum* via gavage (**a**) or intravenous (**b**) routes. H&E-stained sections of colon tissues from gavage-treated (**c**) or intravenously treated (**d**) mice

syringe, taking care to avoid excessive pressure. Transfer the cleaned colon into a 10-cm dish containing cold PBS (*see* **Note 11**).

6. For imaging, in a biosafety cabinet, open the colon longitudinally with fine scissors. Photograph the tumors with a ruler for scale, and then count the number of lesions.
7. For histological analysis, immediately flush the colon with 5–10 mL of formalin, and then place the tissue into a 15-mL tube containing 10 mL of formalin. Fix the sample overnight at room temperature on a slow shaker, followed by immersion in 70% ethanol overnight. Transfer tissues into clearly labeled embedding cassettes and store in 70% ethanol. Submit the fixed tissues for paraffin embedding, sectioning, and H&E staining through institutional pathology services.

3.4 FISH Analysis

Fluorescence in situ hybridization (FISH) provides a highly specific and spatially resolved method for detecting *F. nucleatum* within tissue sections. In this protocol, oligonucleotide probes targeting *F. nucleatum*-specific 16S rRNA sequences enable precise identification of the bacterium within complex microbial communities while preserving native colonic architecture. This approach allows single-cell-level visualization and detailed assessment of *F. nucleatum* localization relative to host structures (*see* Fig. 4).

3.4.1 Slide Pretreatment

1. Immerse slides in xylene at room temperature (RT) for 10 min. Repeat twice, using fresh xylene each time.
2. Rehydrate slides sequentially in 100%, 90%, 70%, and 50% ethanol for 2 min each at RT, and then immerse in water.

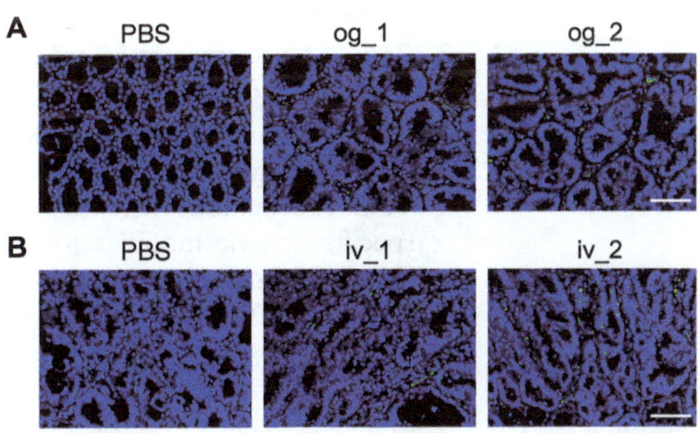

Fig. 4 FISH detection of *F. nucleatum* in colon tumor sections. Cy2-labeled *F. nucleatum* (green; single rods or aggregates) is shown in relation to DAPI-stained host nuclei (blue) in PBS-treated and *F. nucleatum*-treated mice via gavage (top) or intravenous (bottom) administration. Scale bar: 50 μm

3. Apply proteinase solution and incubate slides at 37 °C for 30 min.
4. Dehydrate slides sequentially in 70%, 85%, 95%, and 100% ethanol for 2 min each at RT, and then air-dry.

3.4.2 Denaturation and Hybridization

1. Prepare the hybridization buffer and apply 0.5 mL of probe-containing solution to each hybridization area.
2. Incubate samples with FISH Hybridization Solution in a pre-heated chamber at 95 °C, and then reduce the temperature to 46 °C, and incubate for 2–4 h (*see* **Note 12**).
3. Wash slides with FISH Wash Buffer at 48 °C for 15 min.
4. Dip slides in cold water to remove residual salts, and then air-dry. Keep slides protected from light.
5. Apply DAPI counterstain, add coverslips, and store in the dark.
6. Proceed with imaging and analysis.

4 Notes

1. *F. nucleatum* is highly oxygen-sensitive; even brief exposure to ambient air during handling or transfers can result in significant bacterial death and reduced infection efficiency. Prepare all tubes and media inside an anaerobic chamber, and seal them tightly before removal. Pre-reducing media in the chamber overnight is recommended to maintain optimal anaerobic conditions.
2. Take care to prevent contamination during DNA isolation and PCR. Perform all procedures in a biosafety cabinet, use filter tips, and dedicate a separate set of pipettes exclusively for genotyping.
3. Boiling tail samples in 50 mM of NaOH provides a rapid, cost-effective, and reliable method for releasing genomic DNA.
4. Using a homogeneous genetic background is essential when comparing genetically modified mice with littermate controls. Tumor formation efficiency in the colon is influenced by strain-specific genetic modifier loci. Additionally, diet has a major impact on tumor susceptibility and must remain consistent throughout the experimental period.
5. Prolonged vortexing is required to ensure complete dissolution of tamoxifen.
6. Tamoxifen is light-sensitive and should be stored at −20 °C up to 1 month.

7. The amount of tamoxifen administered directly influences tumor number. A dosage of 2 mg per 20–25 g mice, given twice, produces the maximal tumor burden.
8. If resistance is encountered during gavage, do not force the needle. Withdraw and reposition before attempting again.
9. Ensure syringes are prepared without air bubbles prior to intravenous injection.
10. Avoid rupturing or cutting the intestinal wall, as leakage will prevent effective flushing of fecal material with PBS and formalin.
11. If flushing is unsuccessful due to a puncture in the intestinal wall, identify the site by observing turbulent PBS flow in the dish. Insert the syringe into the hole, and continue flushing until the lumen is fully cleared of fecal material.
12. During FISH, cool samples gradually in the hybridization oven over 2–4 h. Slow cooling promotes efficient annealing of complementary strands, enhances probe specificity, and minimizes secondary structure formation.

Acknowledgments

Work presented in this publication was supported by the National Institute of Dental and Craniofacial Research (NIDCR) of the National Institutes of Health (NIH) under the award number DE033900 (to H.T.-T.).

Conflict of Interest The authors have no conflicts of interest to declare that are relevant to the content of this chapter.

References

1. Bray F, Laversanne M, Sung H, Ferlay J, Siegel RL, Soerjomataram I, Jemal A (2024) Global cancer statistics 2022: GLOBOCAN estimates of incidence and mortality worldwide for 36 cancers in 185 countries. CA Cancer J Clin 74:229–263. https://doi.org/10.3322/caac.21834
2. Avril M, DePaolo RW (2021) "Driver-passenger" bacteria and their metabolites in the pathogenesis of colorectal cancer. Gut Microbes 13:1941710. https://doi.org/10.1080/19490976.2021.1941710
3. Kostic AD, Gevers D, Pedamallu CS, Michaud M, Duke F, Earl AM, Ojesina AI, Jung J, Bass AJ, Tabernero J, Baselga J, Liu C, Shivdasani RA, Ogino S, Birren BW, Huttenhower C, Garrett WS, Meyerson M (2012) Genomic analysis identifies association of *Fusobacterium* with colorectal carcinoma. Genome Res 22:292–298. https://doi.org/10.1101/gr.126573.111
4. Dai Z, Coker OO, Nakatsu G, Wu WKK, Zhao L, Chen Z, Chan FKL, Kristiansen K, Sung JJY, Wong SH, Yu J (2018) Multi-cohort analysis of colorectal cancer metagenome identified altered bacteria across populations and universal bacterial markers. Microbiome 6:70. https://doi.org/10.1186/s40168-018-0451-2
5. Cavallucci V, Palucci I, Fidaleo M, Mercuri A, Masi L, Emoli V, Bianchetti G, Fiori ME, Bachrach G, Scaldaferri F, Maulucci G,

Delogu G, Pani G (2022) Proinflammatory and cancer-promoting pathobiont *Fusobacterium nucleatum* directly targets colorectal cancer stem cells. Biomolecules 12. https://doi.org/10.3390/biom12091256

6. Chen Y, Lu Y, Ke Y, Li Y (2019) Prognostic impact of the *Fusobacterium nucleatum* status in colorectal cancers. Medicine (Baltimore) 98: e17221. https://doi.org/10.1097/MD.0000000000017221
7. Dong J, Li Y, Xiao H, Zhang S, Wang B, Wang H, Li Y, Fan S, Cui M (2021) Oral microbiota affects the efficacy and prognosis of radiotherapy for colorectal cancer in mouse models. Cell Rep 37:109886. https://doi.org/10.1016/j.celrep.2021.109886
8. Li X, Huang J, Yu T, Fang X, Lou L, Xin S, Ji L, Jiang F, Lou Y (2020) *Fusobacterium nucleatum* promotes the progression of colorectal cancer through Cdk5-activated Wnt/beta-catenin signaling. Front Microbiol 11: 545251. https://doi.org/10.3389/fmicb.2020.545251
9. Zheng X, Liu R, Zhou C, Yu H, Luo W, Zhu J, Liu J, Zhang Z, Xie N, Peng X, Xu X, Cheng L, Yuan Q, Huang C, Zhou X (2021) ANGPTL4-mediated promotion of glycolysis facilitates the colonization of *Fusobacterium nucleatum* in colorectal cancer. Cancer Res 81:6157–6170. https://doi.org/10.1158/0008-5472.CAN-21-2273
10. Brennan CA, Nakatsu G, Gallini Comeau CA, Drew DA, Glickman JN, Schoen RE, Chan AT, Garrett WS (2021) Aspirin modulation of the colorectal cancer-associated microbe *Fusobacterium nucleatum*. mBio 12. https://doi.org/10.1128/mBio.00547-21
11. Guo P, Tian Z, Kong X, Yang L, Shan X, Dong B, Ding X, Jing X, Jiang C, Jiang N, Yu Y (2020) FadA promotes DNA damage and progression of *Fusobacterium nucleatum*-induced colorectal cancer through up-regulation of chk2. J Exp Clin Cancer Res 39:202. https://doi.org/10.1186/s13046-020-01677-w
12. Han YW (2015) *Fusobacterium nucleatum*: a commensal-turned pathogen. Curr Opin Microbiol 23:141–147. https://doi.org/10.1016/j.mib.2014.11.013
13. Bell HN, Rebernick RJ, Goyert J, Singhal R, Kuljanin M, Kerk SA, Huang W, Das NK, Andren A, Solanki S, Miller SL, Todd PK, Fearon ER, Lyssiotis CA, Gygi SP, Mancias JD, Shah YM (2021) Reuterin in the healthy gut microbiome suppresses colorectal cancer growth through altering redox balance. Cancer Cell. https://doi.org/10.1016/j.ccell.2021.12.001

Chapter 20

Caenorhabditis elegans as a Model Host for *Fusobacterium nucleatum*

Suhani B. Bhakta and Yi-Wei Chen

Abstract

Fusobacterium nucleatum is a Gram-negative bacterium that plays a central role in biofilm formation within the oral cavity. While traditionally considered a commensal organism, accumulating evidence has implicated *F. nucleatum* in the pathogenesis of a range of oral and systemic diseases, including various cancers, inflammatory bowel disease, and adverse pregnancy outcomes such as preterm birth. Despite its emerging recognition as a pathobiont, relatively few virulence factors have been characterized. In this study, we present *Caenorhabditis elegans* as a versatile and tractable model for investigating *F. nucleatum* virulence. We detail protocols for synchronizing *C. elegans* populations, assessing nematode survival postinfection, and quantifying worm motility to evaluate the physiological impact of infection. These methodologies enable high-throughput screening of *F. nucleatum* virulence determinants and offer a valuable platform for advancing our understanding of its pathogenic potential.

Key words *Fusobacterium nucleatum*, *Caenorhabditis elegans*, Survival, Motility, Pathogenesis

1 Introduction

Fusobacterium nucleatum is a Gram-negative anaerobic bacterium commonly residing in the oral cavity, where it plays a pivotal role in bridging early and late colonizers during dental biofilm formation [1, 2]. While traditionally considered a commensal organism contributing to oral homeostasis, a growing body of evidence has implicated *F. nucleatum* in a range of extraoral diseases [3]. Notably, it has been found enriched in various cancers [4, 5], gastrointestinal disorders [6], and adverse pregnancy outcomes [7, 8]. Despite its expanding clinical relevance, our understanding of the specific fusobacterial genes responsible for extraoral pathogenesis, tissue tropism, and oncogenic activity remains limited. This knowledge gap underscores the urgent need for scalable screening platforms to identify the molecular drivers of

Hung Ton-That (ed.), *Fusobacterium nucleatum: Methods and Protocols*, Methods in Molecular Biology, vol. 3055,
https://doi.org/10.1007/978-1-0716-5388-3_20,

F. nucleatum-mediated colonization, immune modulation, and tumor-promoting activity.

Animal models have been instrumental in elucidating the pathogenic mechanisms of *Fusobacterium nucleatum*, particularly in contexts such as colorectal cancer and preterm birth [9–12]. Among these, murine models have been favored due to their physiological relevance and the ability of *F. nucleatum* to colonize various tissues [13, 14]. However, mouse-based studies present several challenges, including high costs, ethical concerns, and technical limitations that hinder their scalability for large-scale screening [15]. These constraints underscore the need for more accessible and efficient model organisms in the study of *F. nucleatum* virulence.

Caenorhabditis elegans, a free-living nematode that naturally consumes bacteria as food, has emerged as a powerful alternative model for studying host-pathogen interactions. Its advantages include a well-characterized innate immune system, genetic tractability, rapid life cycle, ease of maintenance, and compatibility with high-throughput assays [16–18]. The *C. elegans* infection model has been successfully used to identify virulence factors in diverse pathogens, including pathogenic *Escherichia coli* [19, 20], *Pseudomonas aeruginosa* [21], and *Salmonella enterica* [22, 23]. Building on this foundation, we established and optimized protocols [24, 25] to utilize *C. elegans* as a model host for evaluating *F. nucleatum* virulence. These methods include synchronized infection assays and quantitative assessments of nematode survival and motility, providing a scalable and cost-effective platform for dissecting the pathogenic potential of *F. nucleatum*.

2 Materials

2.1 Preparation of F. nucleatum Nematode Plates

1. *E. coli* OP50 (*Caenorhabditis* Genetics Center; CGC).
2. *F. nucleatum* ATCC 23726.
3. TSPC medium: mix 1X Tryptic Soy Broth with 1% Bacto Peptone and autoclave at 121 °C for 20 min. Before bacterial inoculation, add freshly autoclaved cysteine solution to achieve a final concentration of 0.05%.
4. 37 °C incubator; the incubator can be placed inside an anaerobic chamber maintained with a gas mixture of 95% N_2, 2.5% H_2, and 2.5% CO_2.
6. Worm phosphate buffer: dissolve 119.35 g of KH_2PO_4 and 21.43 g of K_2HPO_4 in deionized water and bring the final volume to 1 L. Adjust the pH to 6. Autoclave the buffer at 121 °C for 20 min.

7. NGM+ agar plate: mix 3 g of NaCl, 10 g of Bacto Peptone, 1 g of yeast extract, 20 g of agar, and 970 mL of deionized water, and then autoclave the medium at 121 °C for 20 min. Once the medium is cooled to 55 °C, add 1 mL of autoclaved 1 M of $CaCl_2$, 1 mL of 1 M $MgSO_4$, 1 mL of 5% cholesterol dissolved in ethanol, and 25 mL of worm phosphate buffer, and swirl to mix well. Pour 10 mL of the medium into each 6-cm petri dish.
8. 6-cm petri dishes.
9. Tips, serological pipettes, microtubes, and centrifuge tubes.
10. Pipettors and pipette aids.

2.2 Synchronization of Worms

1. *C. elegans* N2 strain (gravid adult).
2. Eyelash pickers.
3. *E. coli* OP50 on NGM+ agar plate (*see* Subheading 2.1).
4. Stereomicroscope.
5. 20 °C incubator.

2.3 Nematode Survival and Motility Assays

1. *C. elegans* N2 strain (L4).
2. 20 °C incubator.
3. *E. coli* OP50 on NGM+ agar plate (*see* Subheading 2.1).
4. *F. nucleatum* ATCC 23726 on NGM+ agar plate (*see* Subheading 2.1).
5. Eyelash pickers.
6. Stereomicroscope with CCD.
7. ImageJ.

3 Methods

3.1 Preparation of F. nucleatum Nematode Plate

1. Inoculate individual cultures of *E. coli* OP50 and *F. nucleatum* ATCC 23726 strains into 5 mL of TSPC medium at 37 °C in an anaerobic chamber for 16 to 18 h.
2. Measure the optical density of each bacterial culture at 600 nm (OD_{600}).
3. Normalize the cultures to an OD_{600} of approximately 1.0 using fresh TSPC medium.
4. Spot and evenly spread 300 μL of the normalized bacterial suspensions onto separate 6-cm NGM+ agar plates.
5. Incubate the plates overnight at 37 °C in an anaerobic incubator (gas composition: 95% N_2, 2.5% H_2, and 2.5% CO_2) to allow bacteria lawn formation.

6. Store the prepared plates at room temperature in an anaerobic chamber until use (*see* **Note 1**).

3.2 Synchronization of Worms

1. Using an eyelash picker under a stereomicroscope, transfer approximately 100 gravid adult-stage *C. elegans* (N2 strain) to an NGM+ agar plate seeded with an *E. coli* OP50 lawn.
2. Incubate the plate at 20 °C for 2 h to allow egg-laying. After incubation, carefully remove the adult worms using the eyelash picker.
3. Continue incubating the plate at 20 °C for approximately 50 h. By this time, the eggs will have developed into L4-stage larvae, which are suitable for subsequent assays.

3.3 Nematode Survival Assay

1. Using an eyelash picker under a stereomicroscope, transfer 30 L4-stage *C. elegans* (N2 strain) from the synchronized population (*see* Subheading 3.2) to NGM+ agar plates seeded with either *E. coli* OP50 or *F. nucleatum* ATCC 23726 (prepared in Subheading 3.1) (*see* **Note 2**).
2. Incubate the plates at 20 °C in an aerobic incubator overnight (*see* **Note 3**).
3. The following day, count and record the number of live worms. Transfer the live worms to a fresh plate with the same bacterial lawn.
4. The next day, count and record the number of live worms. Transfer the live worms to a fresh plate with the same bacterial lawn.
5. Repeat steps 2–4 daily until all worms have died, marking the endpoint of the assay.
6. Calculate the daily survival rate (%) using the formula:

   Survival rate (%) = (Number of live worms / 30) × 100%.

   (*See* Fig. 1 for graphical representation.)

3.4 Nematode Motility Assay

1. Using an eyelash picker under a stereomicroscope, transfer ten L4-stage *C. elegans* (N2 strain) from the synchronized population (*see* Subheading 3.2) to NGM+ agar plates seeded with either *E. coli* OP50 or *F. nucleatum* ATCC 23726 (prepared in Subheading 3.1).
2. Ten minutes after transferring the worms, individually video record each worm using the CCD camera attached to the stereomicroscope (*see* **Notes 4** and **5**).
3. Incubate the plates at 20 °C overnight (*see* **Note 3**).
4. Repeat steps 2 and 3 daily for five consecutive days.
5. Load the recorded videos into ImageJ (*see* **Note 6**). Set the scale, and then mark the worm's initial position at time 0 s.

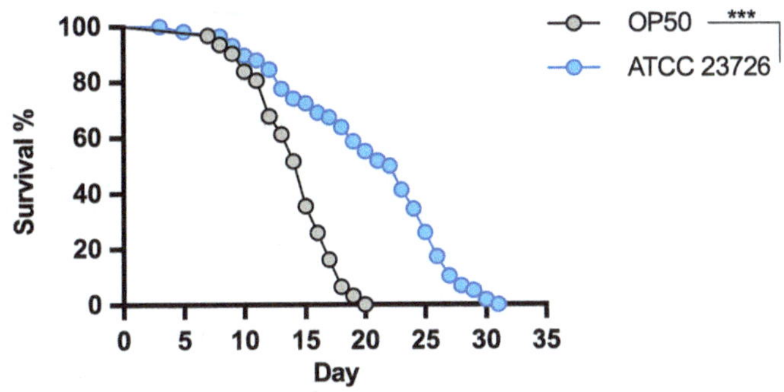

Fig. 1 *F. nucleatum* promotes longevity in *C. elegans*. Thirty L4-stage N2 worms were seeded onto NGM+ agar plate covered with a lawn of *F. nucleatum*. Worms were transferred daily to fresh plates to ensure continuous exposure to fresh bacteria. Survival was monitored daily, and lifespan was calculated accordingly. Statistical analysis was performed using the log-rank (Mantel-Cox) test (***, $P < 0.001$)

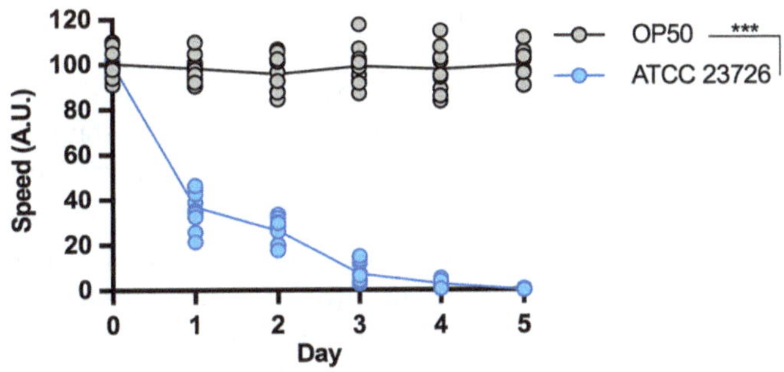

Fig. 2 *F. nucleatum* impairs mobility in *C. elegans*. Ten L4-stage N2 worms were seeded onto NGM+ agar plates with a lawn of *F. nucleatum* following initial video recording of their movement. Worms were transferred daily to fresh plates for continued bacterial exposure, and video recordings were repeated for five consecutive days. Movement distance over time was measured using ImageJ, and worm speed was calculated. Statistical analysis was conducted using one-way ANOVA (***, $P < 0.001$)

Identify the worm's position at approximately the 10-s frame, and apply the initial location to that frame. Record the time (T) to two decimal places.

6. Using the pen tool, trace the worm's movement path to create a curved line.
7. Use the "Measure" function to determine the total distance traveled (D) by each worm.
8. Calculate the moving speed using the formula: $Speed = D/T$ (*see* Fig. 2 for visual presentation).

4 Notes

1. Plates should be used within 3 days of preparation.
2. When setting up and transferring worms during a survival assay, it is important to place the worms directly onto the bacterial lawn rather than on areas without bacteria. This minimizes variability in the initiation of infection and ensures consistency across experimental replicates.
3. Although *E. coli* OP50 is an attenuated strain commonly used as a food source for *C. elegans*, it can outcompete slower-growing anaerobes such as *F. nucleatum* under aerobic conditions. If bacterial contamination is observed along the worms' movement paths after transfer, it is likely due to residual OP50. To address this, an antibiotic wash can be performed prior to transferring worms to experimental plates. Briefly, transfer the worms to an NGM+ plate containing antibiotics that selectively kill OP50 but not *F. nucleatum*—for example, ampicillin or carbenicillin at 50 μg/mL—for 1 h. Then, transfer the worms to the experimental NGM+ plate seeded with *F. nucleatum*. Note that antibiotic resistance may vary among *F. nucleatum* strains; therefore, confirm the strain's susceptibility to the selected antibiotic before initiating the survival assay.
4. A critical aspect of measuring worm motility is ensuring that the worms are in a relatively active and natural state. It is not ideal to record movement when worms are either resting or agitated, as this can introduce variability and reduce the accuracy of speed measurements. To minimize this, we recommend recording worm movement approximately 10 min after transferring them to a new plate—rather than immediately after transfer or following overnight incubation—when their behavior is more stable and representative.
5. Avoid moving the plate during video recording. Once a worm appears near the edge of the view and is likely to move across, stabilize the plate and start recording without interruption to ensure accurate tracking.
6. While ImageJ is recommended here for measuring worm movement, other video or editing software with functions for anchoring positions and displaying timestamps can also be used.

Acknowledgments

We thank Dr. Hung Ton-That for the environmental support and the sharing of resources. We acknowledge the critical comments from the Ton-That laboratory for this manuscript. Work reported in this publication was supported by Innovative Seed Grand of the School of Dentistry at UCLA (to YWC).

Conflict of Interest The authors have no conflicts of interest to declare that are relevant to the content of this chapter.

References

1. Chen Y, Shi T, Li Y, Huang L, Yin D (2022) *Fusobacterium nucleatum*: the opportunistic pathogen of periodontal and peri-implant diseases. Front Microbiol 13:860149. https://doi.org/10.3389/fmicb.2022.860149
2. Mark Welch JL, Rossetti BJ, Rieken CW, Dewhirst FE, Borisy GG (2016) Biogeography of a human oral microbiome at the micron scale. Proc Natl Acad Sci USA 113(6):E791–E800. https://doi.org/10.1073/pnas.1522149113
3. Han YW (2015) *Fusobacterium nucleatum*: a commensal-turned pathogen. Curr Opin Microbiol 23:141–147. https://doi.org/10.1016/j.mib.2014.11.013
4. Wang N, Fang JY (2023) *Fusobacterium nucleatum*, a key pathogenic factor and microbial biomarker for colorectal cancer. Trends Microbiol 31(2):159–172. https://doi.org/10.1016/j.tim.2022.08.010
5. McIlvanna E, Linden GJ, Craig SG, Lundy FT, James JA (2021) *Fusobacterium nucleatum* and oral cancer: a critical review. BMC Cancer 21(1):1212. https://doi.org/10.1186/s12885-021-08903-4
6. Zheng Z, Jin W, Guo W, Jin Z, Zuo Y (2025) Oral *Fusobacterium nucleatum* exacerbates ulcerative colitis via the oral-gut axis: mechanisms and therapeutic implications. Front Cell Infect Microbiol 15:1564169. https://doi.org/10.3389/fcimb.2025.1564169
7. Han YW, Redline RW, Li M, Yin L, Hill GB, McCormick TS (2004) *Fusobacterium nucleatum* induces premature and term stillbirths in pregnant mice: implication of oral bacteria in preterm birth. Infect Immun 72(4):2272–2279
8. Chen YW, Camacho MI, Chen Y, Bhat AH, Chang C, Peluso EA, Wu C, Das A, Ton-That H (2022) Genetic determinants of hydrogen sulfide biosynthesis in *Fusobacterium nucleatum* are required for bacterial fitness, antibiotic sensitivity, and virulence. mBio 13(5):e0193622. https://doi.org/10.1128/mbio.01936-22
9. Yu T, Guo F, Yu Y, Sun T, Ma D, Han J, Qian Y, Kryczek I, Sun D, Nagarsheth N, Chen Y, Chen H, Hong J, Zou W, Fang JY (2017) *Fusobacterium nucleatum* promotes chemoresistance to colorectal cancer by modulating autophagy. Cell 170(3):548–563 e16. https://doi.org/10.1016/j.cell.2017.07.008
10. Han YW, Shen T, Chung P, Buhimschi IA, Buhimschi CS (2009) Uncultivated bacteria as etiologic agents of intra-amniotic inflammation leading to preterm birth. J Clin Microbiol 47(1):38–47. https://doi.org/10.1128/JCM.01206-08
11. Britton TA, Wu C, Chen YW, Franklin D, Chen Y, Camacho MI, Luong TT, Das A, Ton-That H (2024) The respiratory enzyme complex Rnf is vital for metabolic adaptation and virulence in *Fusobacterium nucleatum*. MBio 15(1):e0175123. https://doi.org/10.1128/mbio.01751-23
12. Franklin DS, Chen YW, Chen Y, Wittchen M, Agnew A, Luu A, Whitelegge JP, Zhou ZH, Tauch A, Das A, Ton-That H (2024) Ethanolamine-induced assembly of microcompartments is required for *Fusobacterium nucleatum* virulence. mBio. https://doi.org/10.1128/mbio.03405-24
13. Rydell-Tormanen K, Johnson JR (2019) The applicability of mouse models to the study of human disease. Methods Mol Biol 1940:3–22. https://doi.org/10.1007/978-1-4939-9086-3_1
14. Perlman RL (2016) Mouse models of human disease: an evolutionary perspective. Evol Med Public Health 2016(1):170–176. https://doi.org/10.1093/emph/eow014
15. Mukherjee P, Roy S, Ghosh D, Nandi SK (2022) Role of animal models in biomedical research: a review. Lab Anim Res 38(1):18. https://doi.org/10.1186/s42826-022-00128-1
16. Ermolaeva MA, Schumacher B (2014) Insights from the worm: the *C. elegans* model for innate immunity. Semin Immunol 26(4):303–309. https://doi.org/10.1016/j.smim.2014.04.005

17. Tran TD, Luallen RJ (2024) An organismal understanding of *C. elegans* innate immune responses, from pathogen recognition to multigenerational resistance. Semin Cell Dev Biol 154(Pt A):77–84. https://doi.org/10.1016/j.semcdb.2023.03.005
18. Kaletta T, Hengartner MO (2006) Finding function in novel targets: *C. elegans* as a model organism. Nat Rev Drug Discov 5(5): 387–398. https://doi.org/10.1038/nrd2031
19. Huang CR, Kuo CJ, Huang CW, Chen YT, Liu BY, Lee CT, Chen PL, Chang WT, Chen YW, Lee TM, Hsieh HC, Chen CS (2021) Host CDK-1 and formin mediate microvillar effacement induced by enterohemorrhagic *Escherichia coli*. Nat Commun 12(1):90. https://doi.org/10.1038/s41467-020-20355-1
20. Hashimoto M, Ma YF, Wang ST, Chen CS, Teng CH (2021) Iron Acquisition of Urinary Tract Infection *Escherichia coli* involves pathogenicity in *Caenorhabditis elegans*. Microorganisms 9(2). https://doi.org/10.3390/microorganisms9020310
21. Tan MW, Mahajan-Miklos S, Ausubel FM (1999) Killing of *Caenorhabditis elegans* by *Pseudomonas aeruginosa* used to model mammalian bacterial pathogenesis. Proc Natl Acad Sci USA 96(2):715–720. https://doi.org/10.1073/pnas.96.2.715
22. Aballay A, Yorgey P, Ausubel FM (2000) *Salmonella typhimurium* proliferates and establishes a persistent infection in the intestine of *Caenorhabditis elegans*. Curr Biol 10(23): 1539–1542. https://doi.org/10.1016/s0960-9822(00)00830-7
23. Paulander W, Pennhag A, Andersson DI, Maisnier-Patin S (2007) *Caenorhabditis elegans* as a model to determine fitness of antibiotic-resistant *Salmonella enterica* serovar *typhimurium*. Antimicrob Agents Chemother 51(2):766–769. https://doi.org/10.1128/AAC.00615-06
24. Chen YW, Ton-That H (2020) *Corynebacterium diphtheriae* virulence analyses using a *Caenorhabditis elegans* model. Curr Protoc Microbiol 58(1):e109. https://doi.org/10.1002/cpmc.109
25. Chen YW, Ko WC, Chen CS, Chen PL (2018) Evaluating virulence and pathogenesis of *Aeromonas* infection in a *Caenorhabditis elegans* model. J Vis Exp. https://doi.org/10.3791/58768

FSC
www.fsc.org
MIX
Papier | Fördert
gute Waldnutzung
FSC® C083411

Kolibri 360 GmbH
Ferdinand-Jühlke-Straße 7
99095 Erfurt, Deutschland
produktsicherheit@kolibri360.de